WIRELESS NETWORK DEPLOYMENTS

THE KLUWER INTERNATIONAL SERIES
IN ENGINEERING AND COMPUTER SCIENCE

WIRELESS NETWORK DEPLOYMENTS

edited by

Rajamani Ganesh
GTE Laboratories

Kaveh Pahlavan
Worcester Polytechnic Institute

Kluwer Academic Publishers
Boston/Dordrecht/London

Distributors for North, Central and South America:
Kluwer Academic Publishers
101 Philip Drive
Assinippi Park
Norwell, Massachusetts 02061 USA
Telephone (781) 871-6600
Fax (781) 871-6528
E-Mail <kluwer@wkap.com>

Distributors for all other countries:
Kluwer Academic Publishers Group
Distribution Centre
Post Office Box 322
3300 AH Dordrecht, THE NETHERLANDS
Telephone 31 78 6392 392
Fax 31 78 6546 474
E-Mail services@wkap.nl>

Electronic Services <http://www.wkap.nl>

Library of Congress Cataloging-in-Publication

Printed on acid-free paper.

Printed in the United States of America

Contents

PART III: DEPLOYMENT OF TDMA BASED NETWORKS

PART IV: DEPLOYMENT OF WIRELESS DATA NETWORKS

PREFACE

During the past decade, the wireless telecommunication industry's predominant source of income was cellular telephone service. At the start of the new millennium, data services are being perceived as complementing this prosperity. The cellular telephone market has grown exponentially during the past decade, and numerous companies in fierce competition to gain a portion of this growing market have invested heavily to deploy cellular networks. The main investment for deployment of a cellular network is the cost of the infrastructure, which includes the equipment, property, installation, and links connecting the Base Stations (BS). A cellular service provider has to develop a reasonable deployment plan that has a sound financial structure. The overall cost of deployment is proportional to the number of BS sites, and the income derived from the service is proportional to the number of subscribers, which grows in time. Service providers typically start their operation with a minimum number of sites requiring the least initial investment. As the number of subscribers grows, generating a source of income for the service provider, the investment in the infrastructure is increased to improve the service and capacity of the network to accept additional subscribers. A number of techniques have evolved to support the growth and expansion of cellular networks. These techniques involve methodologies to increase reuse efficiency, capacity, and coverage while maintaining the target quality of service (QoS) available to the subscriber.

Most of the available literature on wireless networks focusses on wireless access techniques, modem design technologies, radio propagation modeling, and design of efficient protocols for reliable wireless communications. These issues are related to the efficiency of the air interface to optimize the usage of the available bandwidth and to minimize the consumption of power, consequently extending the lifetime of the batteries. An important aspect of wireless networks that has not received adequate attention is the deployment of the infrastructure. Most textbooks discuss the abstract mathematics employed in determining frequency reuse factors or the methodologies used in predicting radio propagation to determine the coverage of a radio system. The real issues faced in network deployments, which limit the theoretical capacity, coverage, voice quality, etc., or performance enhancements that take into account the current infrastructure, are not treated adequately. The objective of this book is to address this gap.

To visualize the complexity of a "green field" or an "overlay" deployment, one should first realize that (1) a wireless service provider's largest investment is the cost of the physical site location (antenna, property, and maintenance), and (2) the deployment is an evolutionary process. The service provider starts with an available and potentially promising technology and a minimum number of sites to provide basic coverage to high-traffic areas. To support an increasing number of subscribers, a demand for increased capacity and better quality of service, the service provider also explores use of more sophisticated antennas (sectored or smart), use of more efficient wireless access methods (TDMA or CDMA), and increasing the number of deployed sites and carriers. As a result, in addition to supporting the continual growth of user traffic with time, the service provider needs to be concerned about the impact of changes in the antenna, access technique, or number of sites on the overall efficiency and return on investment of the deployed network. All major service providers have a group or a division equipped with sophisticated and expensive deployment tools and measurement apparatus to cope with these continual enhancements made in the overall structure of the network.

In this book, we have invited a number of experts to write on a variety of topics associated with deployment of digital wireless networks. We have divided these topics into four categories, each constituting a part of the book. The first part, consisting of three chapters, provides an overview of deployment issues. Saleh Faruque of Metricom provides a step-by-step process for system design and engineering integration required in various stages of deployment. Jay Weitzen and Mark Wallace of NextWave Telecom address and compare the issues related to deployment of polarization diversity antenna systems with deployment of the classic two-antenna space diversity system. Michael Zhao, Yonghai Gu, Scott Gordon, and Martin Feuerstein of Metawave Communications Corp. examine the performance of deploying smart antenna architectures in cellular and PCS networks.

The next three parts of the book cover issues involved in deployment of CDMA, TDMA, and Wireless Data networks. The three chapters in Part II concern deployment of CDMA networks based on the IS-95 standard. Part II begins with a chapter by Vincent O'Byrne, Haris Stellakis, and Rajamani Ganesh of GTE that addresses the complex optimization of dual mode CDMA networks deployed in an overlaid manner over the legacy analog AMPS system. The second chapter, by Jin Yang of Vodafone AirTouch, discusses issues related to embedding a microcell to improve hot-spot capacity and dead-spot coverage in an existing macrocellular CDMA network. The last chapter in Part II, by Steven Gray and Giridhar Mandyam of Nokia Research Center in Texas, addresses detection and mitigation of intermodulation distortion in CDMA handset transceivers.

Part III deals with issues found in deployment of TDMA based networks. The first chapter, by Jerome Brouet, Vinod Kumar, and Armelle Wautier of Alcatel and Ecole Supérieure d'Electricité in France, develops the principle of hierarchical systems to meet the traffic demand in high density hot-spots and compares this technique with conventional methods used to enhance the capacity of TDMA networks. The second chapter in Part III, by R. Ramesh and Kumar Balachandran of Ericsson, derives a strategy to maximize the number of ANSI-136 users supported for a given number of AMPS users and considers reconfigurable transceivers at the base station to increase traffic capacity in a dual mode ANSI-136/AMPS network. The last chapter in Part III, by Anwar Bajwa of Camber Systemics Limited in UK, addresses the practical deployment of the frequency hopping feature in GSM networks to realize increased capacity with marginal degradation in QoS.

The final part, Part IV, of this book is devoted to Wireless Data Networks. Wireless data services are divided into (1) mobile data services, providing low data rates (up to a few hundered Kbps) with comprehensive coverage comparable to that of cellular telephones; and (2) Wireless LANs, providing high data rates (more than 1 Mbps) for local coverage and in-building applications. In the first chapter of Part IV, Hakan Inanoglu of Opuswave Network and John Reece and Murat Bilgic of Omnipoint Technologies Inc. discuss fixed deployment considerations of General Packet Radio Services (GPRS) as an upgrade to currently deployed networks and identify system performance for slow-moving and stationary terminal units. The last two chapters deal with deployment of wireless LANs (WLANs). Craig Mathias of Farpoint Group provides an overview of wireless LANs and talks about deployment issues related to placement of access points and interference management. The last chapter, by Anand Prasad, Albert Eikelenboom, Henri Moelard, Ad Kamerman and Neeli Prasad of Lucent Technogies in The

Netherlands, concentrates on coverage, cell planning, power management, security, data rates, interference and coexistence, critical issues for deploying an IEEE 802.11 based WLAN.

We graciously thank all the authors for their contributions and their help with this book, and we hope our readers will find the book's content both unique and beneficial.

Rajamani Ganesh

Kaveh Pahlavan

PART I

OVERVIEW AND ISSUES IN DEPLOYMENTS

Chapter 1

SCIENCE, ENGINEERING AND ART OF CELLULAR NETWORK DEPLOYMENT

SALEH FARUQUE

Metricom Inc.

Abstract: Cellular deployment is a step by step process of system design and system integration which involves, RF Propagation studies and coverage prediction, Identification of Cell site location, Traffic Engineering, Cell planning, Evaluation of C/I etc. In short, it combines science, engineering and art, where a good compromise among all three is the key to the successful implementation and continued healthy operation of cellular communication system. In this paper, we present a brief overview of cellular architecture followed by a comprehensive yet concise engineering process involved in various stages of the design and deployment of the systems.

1. INTRODUCTION

The generic cellular communication system, shown in Fig.1, is an integrated network comprising a land base wire line telephone network and a composite wired-wireless network. The land base network is the traditional telephone system in which all telephone subscribers are connected to a central switching network, commonly known as PSTN (Public Switching Telephone Network). It is a digital switching system, providing: i) Switching, ii) Billing, iii) 911 dialing, iv)1-800 and 1-900 calling features, v) Call waiting, call transfer, conference calling, voice mail etc., vi) Global connectivity. vii) Interfacing with cellular networks. Tens of thousands of simultaneous calls can be handled by means of a single PSTN. The function of the Mobile Switching Center (MSC) or MTX (Mobile Telephone Exchange) is: i) Provide connectivity between PSTN and cellular base stations by means of trunks (T_1 links), ii) Facilitate communication between mobile to mobile, mobile to land, land to mobile and MSC to PSTN, iii) Manage, control and monitor various call processing activities, and iv) Keeps detail record of each call for billing. Cellular base stations are located at different convenient locations within the service area. The coverage of a base station varies from less than a kilometer to tens of kilometers, depending on the propagation environment and traffic density An array of such base stations has the capacity of serving tens of thousands of subscribers in a major metropolitan area. This is the basis of today's cellular telecommunication services.

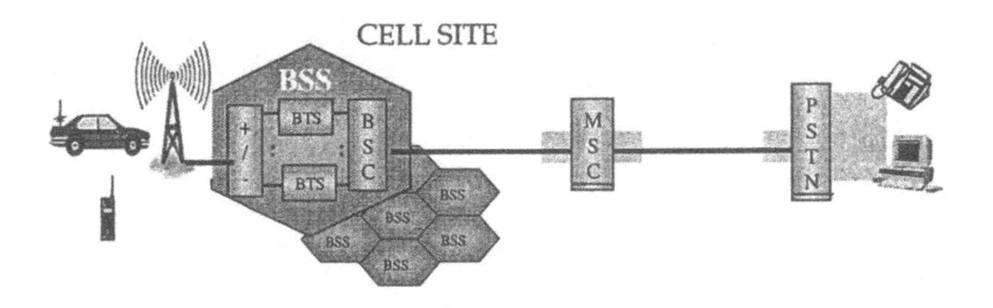

Figure 1. A generic cellular communication network

Cellular deployment, therefore, is a step by step process of system design and system integration involving: a) RF Propagation studies and coverage prediction, b) Cell site location and Tolerance on Cell site Location, c) C/I and Capacity Issues and d) Cell planning. In short, it combines science, engineering and art, where a good compromise among all three is the key to the successful implementation and continued healthy operation of cellular communication system. In this chapter, we present a comprehensive yet concise engineering process involved in various stages of the design and deployment of cellular systems.

2. PROPAGATION ISSUES

Propagation prediction is a process of environmental characterization and propagation studies where the Received Signal Level (RSL) is determined as a function of distance. In a multipath environment, the Received Signal Level is generally chaotic, owing to numerous RF barriers and scattering phenomena which vary from one civil structure to another. Building codes also vary from place to place, requiring wide-ranging databases. Computer aided prediction tools, available today, generally begin with standard propagation models such as Okumura-Hata, Cost-231 or Walfisch-Ikegami model. These models are based on empirical data and their accuracy depends on several variables such as, Terrain elevation data, Clutter factors (Correction factors due to Buildings, Forests, Water etc.), Antenna height, Antenna pattern, ERP (Effective Radiated Power), Traffic distribution pattern, Frequency planning etc. These prediction models are essential during the initial planning, quotation and deployment of cellular communication systems.

Introduction

Radio link design is an engineering process where a hypothetical pathloss is derived out of a set of physical parameters such as ERP, cable loss, antenna gain and various other design parameters. A sample worksheet is then produced for system planning and dimensioning radio equipment. It is a routine procedure in today's mobile cellular communication systems. Unfortunately, the cellular industries have overlooked a potential link between these practices and propagation models they use. As a result the traditional process of link design is generally inaccurate due to anomalies of propagation.

In an effort to alleviate these problems, this section examines the classical Okumura-Hata and the Walfisch-Ikegami models, currently used in land-mobile communication services, and provides a methodology for radio link design based on these models. It is shown that there is a unique set of design parameters associated with each model for which the performance of a given RF link is optimal in a given propagation environment [1].

Classical Propagation Models and their Attributes to Radio Link Design

The classical Okumura-Hata and the Walfisch-Ikegami propagation models exhibit equation of a straight line (Appendix A and B):

$$Lp(dB) = L_o(dB) + 10\,\gamma\log(d) \qquad (1)$$

where Lp is the path loss and Lo is the intercept which depends on antenna height, antenna location, surrounding buildings, diffraction, scattering, road widths etc., γ is the propagation constant or attenuation slope and d is the distance. The parameters Lo and γ are arbitrary constants. These constants do not change once the cell site is in place. Solving for d, we obtain

$$d = 10^{\frac{L_p - L_o}{10\gamma}} \qquad (2)$$

Eq. (2) indicates that there are four operating conditions:

i) The exponent, E, of eq.2 is zero, for which d = 1 and independent of γ (Multipath tolerant).

ii) The exponent of eq.2 is constant for which d > 1 and insensitive to the variation of propagation environment (also multipath tolerant).

iii) The exponent of eq.2 is +ve for which d < 1 and inversely proportional to γ (Multipath attenuation).

iv) The exponent of eq.2 is -ve for which d > 1 and proportional to γ (Multipath gain or wave-guide effect).

These operating conditions are illustrated in Fig.2.

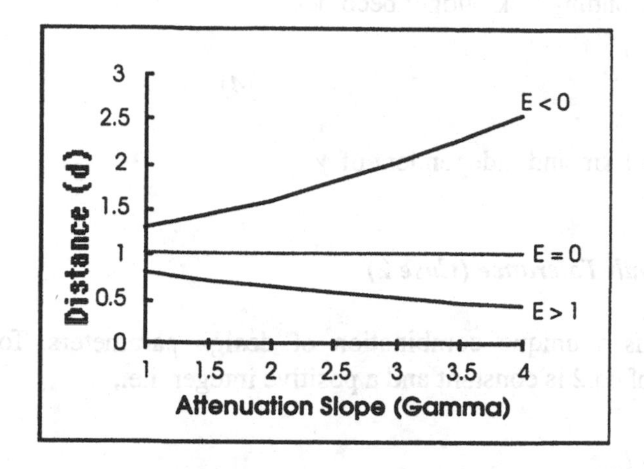

(a)

(b)

Figure 2. Relative coverage as a function of attenuation slope for various operating conditions

The corresponding link budget that satisfies these conditions is as follows:

i) Multipath Tolerance (Case 1)

There is a unique combination of design parameters, for which the exponent of equation 2 vanishes, i.e.,

$$\frac{L_p - L_o}{10\gamma} = 0 \qquad (3)$$

The corresponding link budget becomes

$$L_p = L_o \quad (4)$$

where d = 1 km and independent of γ.

ii) Multipath Tolerance (Case 2)

There is a unique combination of design parameters, for which the exponent of eq.2 is constant and a positive integer, i.e.,

$$\frac{L_p - L_o}{10\gamma} = c \quad (5)$$

for which d > 1 km and independent of γ.

iii) Multipath Attenuation

Multipath attenuation is due to destructive interference where the reflected and diffracted components are > 180deg. Under this condition the link budget can be calculated by setting the exponent of eq. 2 to +ve , i.e,

$$\frac{L_p - L_o}{10\gamma} > 0 \quad (6)$$

for which, d > 1 km but sensitive to γ. Today's cellular communication systems fall largely into this category.

iv) Multipath Gain

Multipath gain is due to constructive interference (wave guide effect), where the reflected and diffracted components are <180 deg. out of phase and form a strong composite signal. Under this condition, the link budget can be calculated by setting the exponent of eq.2 to -ve:

$$\frac{L_p - L_o}{10\gamma} < 0 \quad (7)$$

for which, d < 1 km and sensitive to γ. The path loss slope under this condition is generally < 2, which means that the propagation is better than free space.

It follows that there is a unique set of design parameters for which the average path loss is linear and independent of γ. The radii available in this region is $\leq$ 1 km which is suitable for cellular and μ-cellular services.

3. CELL SITE LOCATION ISSUES

Often, it is not possible to install a cell site in the desired location due to physical restrictions and the cell site has to be relocated, preferably in a nearby location. As a result, the D/R ratio will change, affecting the Carrier to Interference ratio (C/I). In this section we examine the degradation of C/I due to cell site relocation and determine the maximum allowable relocation distance for which C/I = 18 dB.

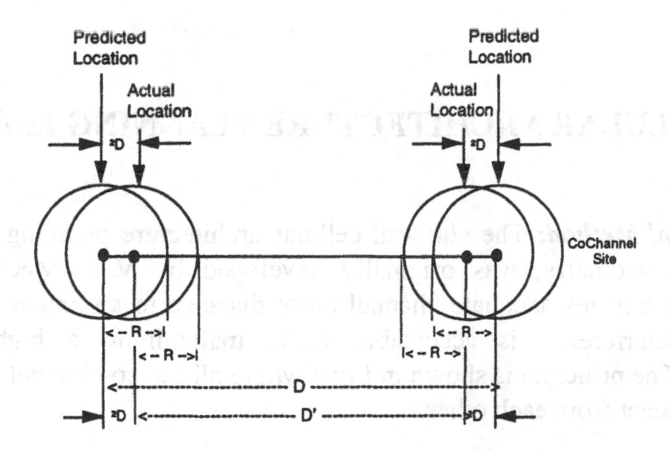

Fig.3 Cell site location tolerance

Consider a pair of co-channel sites having a reuse distance D as shown in Fig. 3. Because of geography and physical restrictions, both cell sites have to be relocated. Let's assume that both cell sites approach each other by ΔD.

The new reuse distance is therefore D-2ΔD. The corresponding C/I then becomes

$$\frac{C}{I} = 10\log\left[\frac{1}{j}\left(\frac{D-2\Delta D}{R}\right)^{\gamma}\right]$$

with

$$\frac{D}{R} = \sqrt{3N}$$

and $\alpha = \Delta D / D$ we obtain

$$\frac{C}{I} = 10\log\left[\frac{1}{J}\left\{\sqrt{3N}(1-2\alpha)\right\}^{\gamma}\right]$$

where J = number of cochannel interferers, N=frequency reuse plan, γ=pathloss slope and $0 < \alpha < 1$. It follows that a number of engineering considerations are involved at this stage before going further. These are (a) C/I vs. Capacity (b) Frequency reuse plan, (c) OMNI vs. Sectorization etc. We discuss some of these issues in the following sections.

4. CELLULAR ARCHITECTURE PLANNING ISSUES

A. Classical Method: The classical cellular architecture planning, based on hexagonal geometry, was originally developed by V.H. MacDonald in 1979[1]. It ensures adequate channel reuse distance to an extent where co-channel interference is acceptable while maintaining a high channel capacity. The principle is shown in Fig.4 where all the co-channel interferers are equidistant from each other.

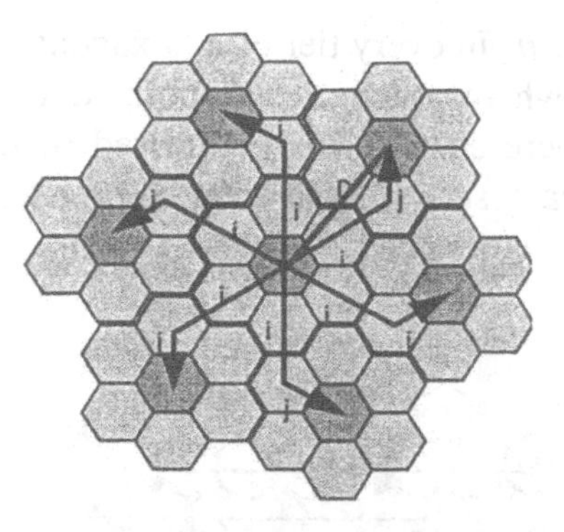

Fig.4. Prior art of N=7 OMNI plan

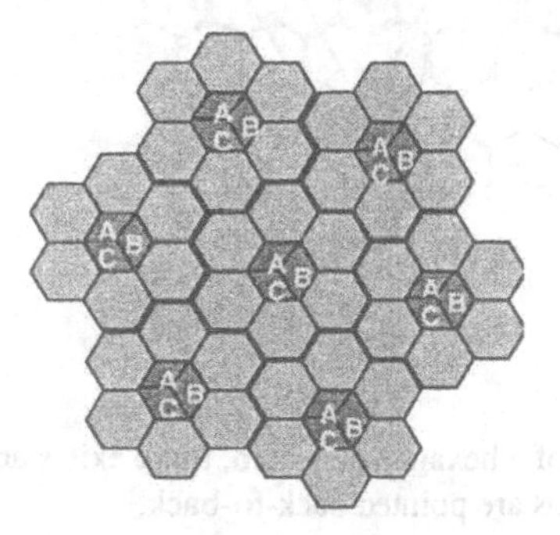

Fig.5. Prior art of N=7 Sectorized plan

This configuration provides a carrier to interference ratio:

$$\frac{C}{I} = 10\log\left[\frac{1}{k}\left(\frac{D}{R}\right)^{r}\right]$$

where $\frac{D}{R} = \sqrt{3N}$ **, D=Frequency reuse distance, R=Cell radius,**

$N = i^2 + ij + j^2$ **, i and j are known as shift parameters, 60^{0} apart and k is the total number of co-channel interferers. In general, k=6 for OMNI plan (Fig.4) and k=3 for tri-sectored plan (Fig.5). From the above illustrations we see that C/I performance depends on two basic parameters: i) Number of interferers and ii) Reuse distance. We also notice that the effective number of interferers is 50% reduced in the 120-degree sectorized system. Yet, there is need to further reduce the C/I interference and enhance capacity.**

B. Directional Reuse Plan: In every tier of a hexagonal system, there exists an apex of a triangle where antennas are pointed back-to-back. This is illustrated in FIG. 6 where each cell is comprised of three sectors having directional antennas in each sector. Each antenna radiates into the respective 120° sector of the three-sectored cell.

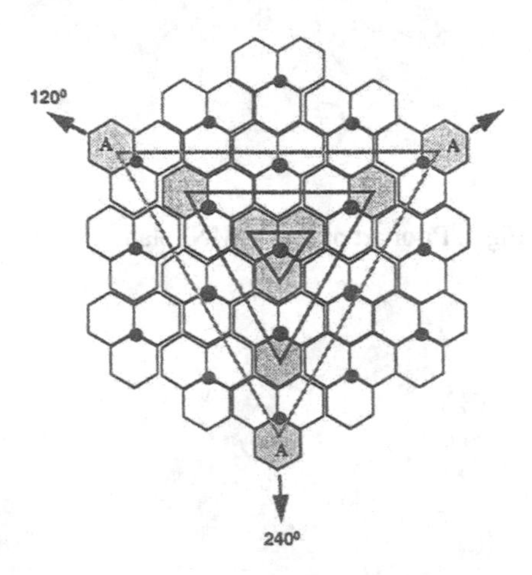

Fig.6. In every tier of a hexagonal system, there exists an apex of a triangle where antennas are pointed back-to-back.

The directional reuse is based on dividing up the available channels into L^2 groups, arranged as an L x L matrix. These L x L matrices are then reused horizontally and vertically according to the following scheme:

[L x L] [L x L] . .
[L x L] [L x L] . .

where $L = 1 + 3i, \quad i = 1, 2, \ldots$

An example of a 4 x 4 array (i = 2, L = 4) shown below, has 16 frequency groups. These groups are arranged alternately to avoid adjacent channel interference. Here, each group has 416/16 = 26 frequencies per group. These groups are then distributed evenly among sectors in a 4 x 4 array according to the following principle:

Cluster of 4 x 4 Array

1	3	5	7	1	3	5	7
9	11	13	15	9	11	13	15
2	4	6	8	2	4	6	8
10	12	14	16	10	12	14	16
1	3	5	7	1	3	5	7
9	11	13	15	9	11	13	15
2	4	6	8	2	4	6	8
10	12	14	16	10	12	14	16

The top 4 x 2 of each array being alternate odd frequency groups and the bottom 4 x 2, alternate even frequency groups. Each frequency group is assigned to a sector according to FIG.6, which automatically generates a back-to-back triangular formation of same frequencies throughout the entire network. The frequency reuse plan as illustrated in FIG.7 is then expanded as needed, in areas surrounding the first use, as required to cover a geographical area. FIG. 6 also illustrates the triangular reuse of frequencies. For example, focusing on frequency group 1, this frequency group is reused after frequency group 7 on the same line as the first use of frequency group 1. Frequency group 1 is also reused at a lower point from the first two uses and such that a triangle is formed when connecting each adjacent frequency group reuse. Each adjacent frequency reuse of the triangle is radiating in a different direction.

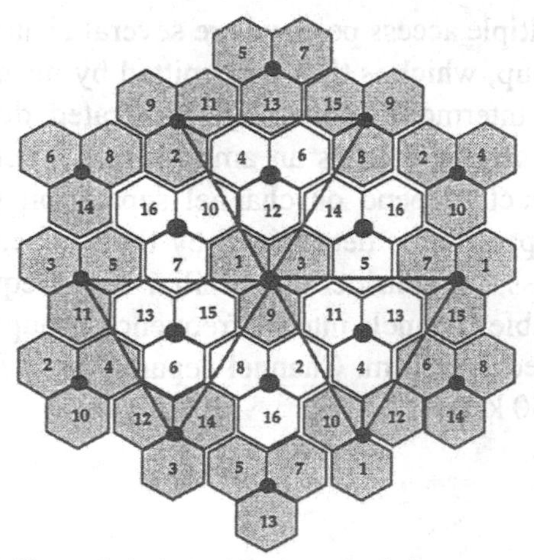

Fig.7. Expanded view of Back-to-Back frequency reuse in three sectored systems

The directional frequency plan of the proposed plan reduces interference such that the effective number of interferers are reduced to two. The C/I of this plan is determined as follows:

$$\frac{C}{I} \approx 10\log\left[\frac{1}{2}\left(\sqrt{3N}\right)^{\gamma}\right] + \Delta dB \approx 21dB$$

where N=5.333, $\gamma = 4$, and ΔdB is due to the antenna side-to-side ratio (> 10 dB for typical sector antennas). The pathloss slope, γ, also referred to in the art as the propagation constant, is the rate of decay of signal strength as a function of distance. This constant is well known in the art and is discussed above. The C/I objective for TDMA is to get a value that is equal to or greater than 18 dB. Obviously, since the present invention provides a C/I of 21 dB, this objective is met. The channel capacity provided by the N=5.333 frequency layout plan of the present invention is determined by dividing the total number of frequency groups, 416, by the number of sectors, 16. In the present case, the frequency layout plan provides $416/16$ = 26 channels per sector.

5. CELLULAR INTERMOD ISSUES

A cell site is a multiple access point where several channels are combined to form a channel group, which is then transmitted by means of the antenna, as shown in Fig. 8. Intermod products are generated during this process through a non-linear device such as an amplifier or a corroded connector. These Intermod products depend on channel separation within the group, where the channel separation is determined by the frequency plan. In order to examine this process, we consider the familiar N=7 frequency plan, based on dividing the available channels into 21 frequency groups, 16 channels per group in non-expanded spectrum. Channel separation_ within this group is given by 21 x 30 = 630 kHz.

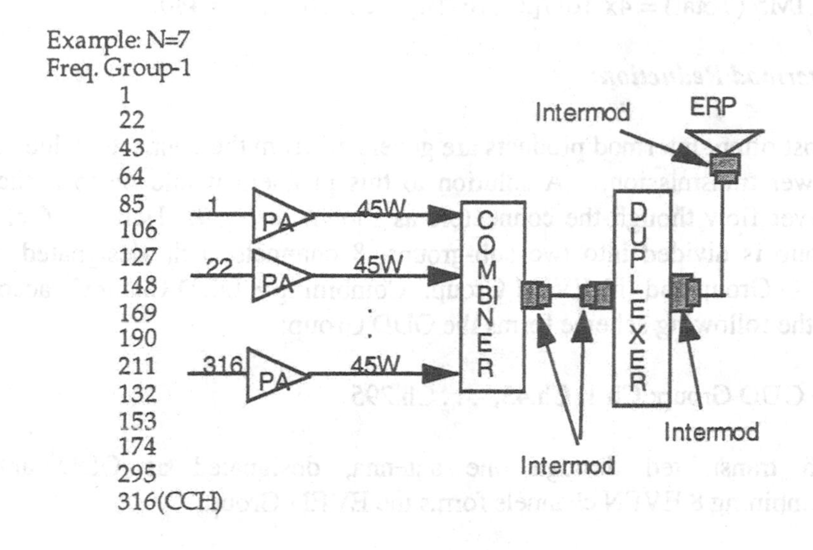

Fig. 8. Origin of Intermod products in a typical cell site

Then a given frequency, say f2 can be related to f1 by means of the following equation:

$$f_2 = f_1 + 0.03 * 21 * k, \text{ where } k = 1,2,\ldots,16,$$

Therefore, the 3rd order intermod products can be written as:

IM3 (in-Band) = $(f_1 - 0.63k)$, $(f1 + 1.26k)$

IM3 (out-of- band)= $(3f1 + 0.63k)$, $(3f1 + 1.26k)$

and the total number of IM3 products due to 16-channel combination appears as:

IM3 (Total) = $4\text{x } 16! / [2!\,(16-2)!] = 2 \text{ x } 16 \text{ x } 15 = 480$.

Similarly, the 5th order intermod products are given by:

IM5 (in-band) = $(f1 - 1.26k)$, $(f1 + 1.89k)$

IM5 (out-of- band) = $(5f1 + 1.26k)$, $(5f1 + 1.89k)$

and the total number of IM5 products due to each 16 channel combinations:

IM5 (Total) = 4x 16! /[2! (16-2)!] = 2 x 16 x 15 = 480.

Intermod Reduction:

Most often, Intermod products are generated from the connectors due to high power transmission. A solution to this problem would be to reduce the power flow though the connectors as shown in Fig.9. Here, a 16-channel group is divided into two sub-groups, 8 channels each, designated as: (i) ODD Group and (ii) EVEN Group. Combining 8 ODD Channels according to the following scheme forms the ODD Group:

ODD Group: Ch.1, Ch.43, ... Ch.295

and transmitted through one antenna, designated as ODD antenna. Combining 8 EVEN channels forms the EVEN Group:

EVEN Group: Ch.22, Ch.64, ... Ch.316

and transmitted through a second antenna, designated as EVEN antenna. This antenna is generally the diversity antenna, which is normally used for space diversity. The effective power flow, as seen by each path, is now reduced by 50%. As a result, the Intermod products are expected to reduce or be virtually eliminated since the slope of 3rd order Intermod power is three times the main power and the slope of 5th order Intermod power is five times the main power. Moreover, the effective channel separation, as seen by each combiner is also increased by a factor of two, i.e. 21 x 2 = 42 (30 kHz x 42 = 1260 kHz), reducing combiner insertion loss. Furthermore, the total number of intermod products are also reduced from 480 per group to 112 per group as shown below:

Number of IM Products in each path = 4x8!/[2!(8-2)!] = 2 x 8 x 7 = 112

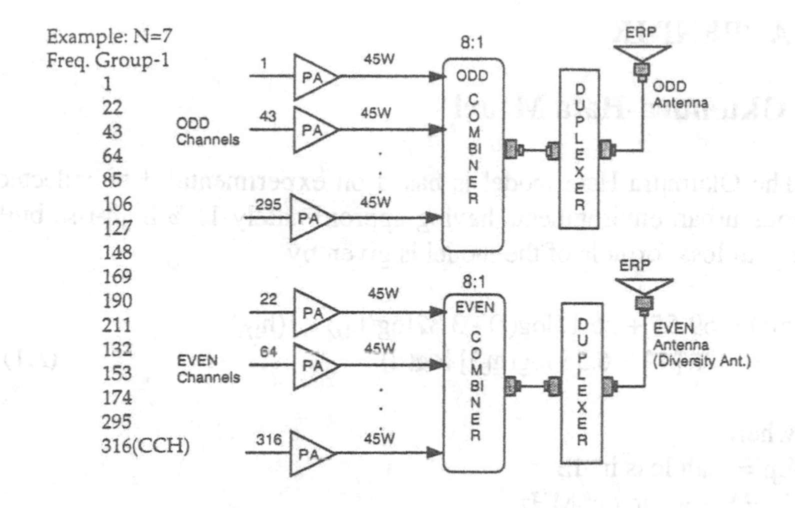

Fig. 9 Method of reducing intermod problems based on antenna sharing which reduces power flow by 50% in each path.

6. CONCLUDING REMARKS

Cellular network deployment is partly science, partly engineering and mostly art. This is due to the fact that RF propagation is "fuzzy" owing to numerous RF barriers and scattering phenomena. Building codes vary from place to place making it practically impossible to rely on software prediction tools. Consequently we end up with drive test, collect data and fine-tune the model. Even then, a margin of 8 to 10 dB in receive signal level is allowed in the final design. These are the realities of RF design with respect to cellular deployment. We have addressed many of these issues in this paper namely, RF propagation, C/I, Frequency planning, cell site location, intermod issues etc. and proposed possible solutions to enhance capacity and performance. If my readers find this information useful, I shall be amply rewarded.

7. APPENDIX

A. Okumura-Hata Model

The Okumura-Hata model is based on experimental data collected from various urban environments having approximately 15% high-rise buildings. The path loss formula of the model is given by

$$L_p(dB) = 69.55 + 26.16\log(f) - 3.82\log(h_b) - a(h_m) + [44.9-6.55 \log(h_b)] \log(d) \quad (A1)$$

where,
Lp = path loss in dB
f = Frequency in MHz
d = Distance between the base station and the mobile(km)
h_b = Effective height of the base station in meters
$a(h_m) = \{1.1 \log(F) - 0.7\}h_m - \{1.56 \log(F) - 0.8\}$
h_m = Mobile antenna height

Eq.(A1) may be expressed conveniently as

$$L_p(dB) = L_o(dB) + [44.9 - 6.55 \log(h_b)] \log(d) \quad (A2)$$

or more conveniently as

$$Lp(dB) = L_o(dB) + 10\gamma \log(d) \quad (A3)$$

where

$$L_o(dB) = 69.55 + 26.16 \log(f) -13.82 \log(h_b) -a(h_m) \quad (A4)$$

and

$$\gamma = [44.9 - 6.55 \log(h_b)]/10 \quad (A5)$$

Eq.(A5) is plotted in Fig.A1 as a function of base station antenna height. It shows that in a typical urban environment the attenuation slope varies between 3.5 and 4.

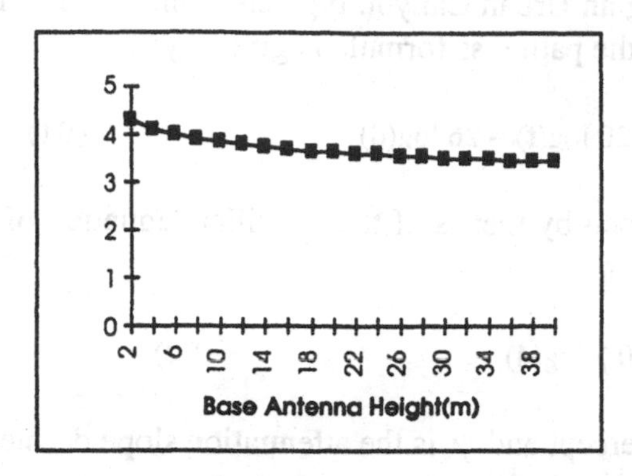

Fig. A1 Attenuation slope as a function of base station antenna height in a typical urban environment (due to Okumura-Hata).

From eq. (A3) we also notice that the Okumura-Hata model exhibits linear path loss characteristics as a function of distance where the attenuation slope is γ and the intercept is L_O. Since L_O is an arbitrary constant, we write

$$L_p(dB) \propto 10\gamma \log(d) \qquad (A6)$$

and in the linear scale,

$$L_p(dB) \propto \frac{1}{d^\gamma} \qquad (A7)$$

(γ = 3.5 to 4)

B. Walfisch-Ikegami Model

The Walfisch-Ikegami model is useful for dense urban environments. This model is based on several urban parameters such as building density, average building height, street widths etc. Antenna height is generally lower than the average building height, so that the signals are guided along the street, simulating an Urban Canyon type environment. For Line Of Sight (LOS) propagation, the path loss formula is given by:

$$L_p(LOS) = 42.6 + 20\log(f) + 26\log(d) \quad \text{(B1)}$$

which can be described by means of the familiar "equation of straight line" as

$$L_p(LOS) = L_o + 10\gamma\log(d) \quad \text{(B2)}$$

where L_o is the intercept and γ is the attenuation slope defined as

$$L_o = 42.6 + 20\log(f)$$
$$\gamma = 2.6$$

Such a low attenuation slope in urban environments (γ=2.6) is believed to be due to low antenna heights (below the rooftop), generating wave guide effects along the street.

For Non Line Of Sight (NLOS) propagation, the path loss formula is

$$L_p(NLOS) = 32.4 + 20\log(f) + 20\log(d) + L(diff) + L(mult) \quad \text{(B3)}$$

where

f , d = Frequency and distance respectively.
L(diff.) = Roof-top diffraction loss
L(mult) = Multiple diffraction loss due to surrounding buildings

The rooftop diffraction loss is characterized as

$$L(diff.) = -16.9 - 10\log(\Delta W) + 10\log(f) + 20\log(\Delta h_m) + L(\varnothing) \quad \text{(B4)}$$

where the parameters in eq.(B4) are defined as

ΔW = distance between the street mobile and the building
h_m = Mobile antenna height
h_{roof} = Average height of surrounding small buildings
$\Delta h_m = h_{roof} - h_m$
$L(\varnothing)$ = Loss due to elevation angle

Multiple diffraction and scattering components are characterized by following equation:

$$L(mult) = k_o + k_a + k_d.\log(d) + k_f.\log(f) - 9\log(W) \quad \text{(B5)}$$

where

$k_o = -18\log(1+\Delta h_b)$

$k_a = 54 - 0.8(\Delta h_b) \quad d \geq 0.5km$

$ = 54 - 0.8(\Delta h_b) \quad d \leq 0.5km$

$k_d = 18 - 15\,(\Delta h_b/h_{roof})$

$k_f = -4 + 0.7[(f/925) - 1]$ for suburban

$ = -4 + 1.5[(f/925) - 1]$ for urban

W = Street width

h_b = Base station antenna height

h_{roof} = Average height of surrounding small buildings ($h_{roof} < h_b$)

$\Delta h_b = h_b - h_{roof}$

It is assumed that the base station antenna height is lower than tall buildings but higher than small buildings.

Combining eq. (B3), eq.(B4) and eq.(B5) we obtain

$$\begin{aligned} L_p(NLOS) &= L_o + (20 + k_d)\log(d) \\ &= L_o + 10\gamma\log(d) \qquad \text{(B6)} \end{aligned}$$

The arbitrary constants are lumped together to obtain

$$\begin{aligned} L_o = {} & 32.4 + (30+k_f)\log(f) - 16.9 - 10\log(w) + 20\log(\Delta h_m) + L(\phi) \\ & + k_o + k_a - 9\log(W) \end{aligned}$$

$$\gamma = (20 + k_d)/10 \quad \text{(B7)}$$

Hence the NLOS characteristics shown in eq.(B6) also exhibits a straight line with L_o as the intercept and γ as the slope. The diffraction constant k_d depends on surrounding building heights, which vary from one urban environment to another, and can vary from a few meters to tens of meters. Typical attenuation slopes in these environments range from $\gamma = 2$ for $\Delta h_b/h_{roof} = 1.2$ to $\gamma = 3.8$ for $\Delta h_b/h_{roof} = 0$. This is shown in Fig.B1.

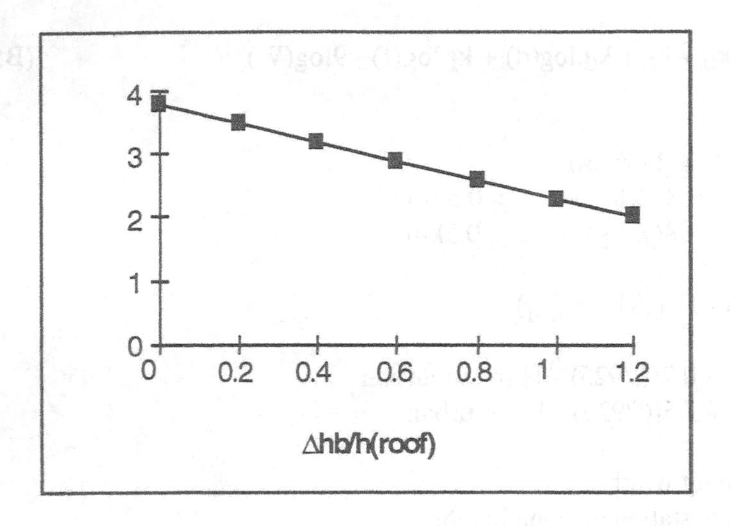

Fig. B1 Attenuation slope as a function of base station antenna height in a typical dense urban environment (due to Walfisch-Ikegami).

REFERENCES

[1] V.H. Mac Donald, "The Cellular Concept", The Bell System Technical Journal", Vol.58, No.1, January 1979.

[2] IS-95 "Mobile Station - Base Station Compatibility Standard for dual Mode Wide band Spread Spectrum Cellular Systems", TR 45, PN-3115, March 15. 1993.

[3] Saleh Faruque, "PCS Micro cells Insensitive to Propagation Medium", IEEE Globecom'94, Proceedings, Vol.1,. pp 32-36.

Chapter 2

COMPARISION OF POLARIZATION AND SPACE DIVERSITY IN OPERATIONAL CELLULAR AND PCS SYSTEMS.

JAY A WEITZEN, MARK S. WALLACE

NextWave Telecom

Abstract: Antenna systems based on polarization diversity can be significantly smaller and easier to deploy than conventional vertically polarized horizontal space diversity systems. As such there is great interest in the substitution of polarization diversity for space diversity. This chapter compares and evaluates the efficacy of polarization diversity relative to the classic vertically polarized 20-wavelength, two-antenna space diversity configuration. It was observed that bottom line performance with a randomly oriented handheld unit was almost identical for polarization and space diversity systems. For a vertical mobile antenna the bottom line performance was approximately 3 dB worse for the polarization diversity system relative to the horizontal space diversity system with vertical polarization. Significant polarization discrimination, which is one slant favored over the other, was observed at close ranges (less than 1 km) when there is a nearly clear line of sight between mobile and base. Significant depolarization was observed at longer ranges and when the mobile is in the clutter.

1. BACKGROUND

Obtaining local zoning authority and other government permission to construct cell sites is one of the most critical paths in the design and operation of a PCS or cellular system. No one wants the big cellular tower in his or her back yards. One of the factors, which make the cell sites so obtrusive, is the large superstructure required for diversity receivers. Diversity reception is used on the uplink in cellular and PCS base stations to combat the effects of multipath induced Rayleigh fading which can cause outages in both analog and digital systems. The theory is that two receivers (antennas) spaced far enough apart will fade independently so that the probability of both receivers simultaneously fading is very low. Various combining techniques including maximal ratio and selection diversity are used depending on the system. Horizontal space diversity using vertical polarized antennas has been the standard configuration for cellular base stations for many years [1,2,3,4,5,6]. The 10 to 20 wavelength horizontal spacing (10 to 20 feet depending on the frequency) between antennas required to achieve a cross-correlation of less than 0.7, drives the design of the large superstructure on cellular towers. This increases both the cost and size of the structure and the difficulty in obtaining permission from local zoning boards to erect new structures. In addition, many landlords now charge by the number of antennas (a total of 6 to 12 per 3-sector cell depending on whether a diplexer is used). The standard horizontal space diversity configuration has been shown to be effective in providing good diversity performance for a subscriber with an antenna mounted vertically on a vehicle. It has also been shown to provide good diversity performance for a user with a randomly oriented handheld portable terminal. Vehicle mounted vertical antennas are being phased out in cellular and are not supported in PCS systems.

For PCS systems that are based on users with handheld portable terminals, polarization diversity can in many circumstances reduce the time, cost and size of the base station antenna array. One quickly observes that the antennas for handheld devices are positioned at random angles, and therefore launch a wave that has significant horizontal and vertically polarized components. The issue comes down to the correlation between the horizontal and vertical components, that is whether there is inherent polarization diversity in the waves launched by hand held portable terminals.

The use of polarization diversity in mobile radio systems is not new [4]. While the use of polarization diversity did not make sense for a system with a large number of vehicle mounted cellular mobile users [2], the rapid increase in the number of hand-held units coupled with increasing difficulties and costs associated with base station deployment in urban areas has lead to a resurgence of interest. Polarization diversity reception systems, at the base stations, which capitalize on the existence of close to equal amplitude signals in two orthogonal components of portable signals, may at the same time provide better performance while reducing the need for the large superstructure.

2. DEFINITION OF DIVERSITY GAIN AND PERFORMANCE MEASURES

For equal amplitude signals in two or more branches, diversity gain is often associated with cross-correlation between branch signals. A cross-correlation of less than 0.7 is generally considered to provide a reasonable improvement in overall performance [4,6]. This is the case with vertical polarized, horizontal space diversity systems. In polarization diversity systems in which the average signal amplitudes may be very different, looking at cross-correlation alone is not an effective measure. For example, if the average signal in two branches of a diversity system differs by 10 dB, even if the signals are uncorrelated, there may not be significant diversity effect. This is an advantage of the 45-degree polarization diversity systems relative to the horizontal/vertical. There is a much greater likelihood that the signals will be balanced, albeit possibly a dB or two lower in some cases, making up for the reduced signal with greater net diversity gain.

A more general method for computing the effective diversity gain was described by Lee and Yeh [4] and was used in the analysis presented here and by other researchers at 800 MHz and at 1.8 GHz [1,2]. The dB level for the 3% cumulative probability (97% reliability) is calculated for a single antenna in the system. Some researchers use 90% and some have used 97%. We have selected the 97% reliability point because of the deleterious effect of deep fades on PCS radio systems and to be consistent with past efforts.

For the CDMA system of interest in the analysis, the next step in the analysis is to form a maximal ratio diversity combined signal by taking, point by point the sum of the power in the two branches. The dB level of the 3% cumulative probability (97% reliability) is calculated for the combined signal. The difference between the 3% cumulative level of the combined signal and the vertically polarized signal is defined as the diversity gain. Implicit in this calculation is the assumption that the signals in the two branches are approximately equal in average amplitude. For identically distributed independent Rayleigh fading signals, the theoretical maximal ratio combining diversity gain at the 3% cumulative probability level is approximately 9.6 dB. This is illustrated in Figure 1 which shows the cumulative distribution functions of the received signal power of a single Rayleigh signal and a 2-branch maximal ratio diversity combined signal formed from two independent Rayleigh distributed signals.

Depending on the amplitudes of the two branches, a diversity gain greater than the 9.6 dB theoretical level is possible, though there is some question as to what it means. If the gains in the two branches are not balanced, with the gain in the reference antenna less than the second antenna, the diversity gain, by definition, will be greater than 9.6 dB and will be dominated by the enhanced signal of the second branch. If the reference branch signal is dominant or the signals are highly correlated, then the reverse is true and the diversity gain is low. Differences in the mean signal level are attributable to cross polarization discrimination at short ranges due to the angle of the transmitting antenna, differences in horizontal versus vertical propagation path loss conditions, or imbalances in the receive antenna patterns.

3. EXPERIMENT DESCRIPTION

Nextwave Telecom conducted a series of experiments to measure the bottom line performance of space diversity relative to polarization diversity to help us decide whether the operational and financial advantages of polarization diversity might be offset by possible performance degradations. A second objective was to assess when and where polarization diversity should and should not be used. Four spectrum analyzers were used as calibrated narrow band receivers. Low noise amplifiers with about 22 dB gain (powered off the probe port of the analyzers) were used at the front end of the spectrum analyzers to improve the overall noise figure to about 5 dB. The analyzers were phase locked and set to the zero span mode using a 3 kHz RF bandwidth and a 1 kHz video bandwidth. The video output of the analyzers was connected to a multi-channel twelve-bit A/D converter logging data at a rate of 2000 samples per second per channel. The high logging speed allows observation of the Rayleigh fading component of the signals. Each measurement system was carefully calibrated to compensate for slight differences in cable losses and preamplifier gain.

Calibration of input signal level vs. output voltage was performed at the beginning and end of the experiment. In the measurement systems 0.01-Volt change in the video output level represents approximately 1 dB of signal level change. The accuracy and stability of the calibration and therefore the experiment is approximately plus or minus 1 dB.

Continuous wave (CW) signals were transmitted from the vehicle to the receive test site atop the 19th floor of the Fox Hall dormitory at University of Massachusetts Lowell, approximately 210 feet above ground level. This site is the tallest building in Northern Middlesex County and has a view to a variety of morphology types including light urban (Downtown Lowell with closely spaced 5-8 story buildings) suburban residential with trees, and open residential. The terrain is relatively flat to gently rolling within the coverage region of the receiver.

A set of drive routes was selected in Lowell, Massachusetts on the boresight of the antenna and to the sides of the 90 horizontal pattern. Drive routes were selected to provide a sample of morphologies and distances from the base station and are described in Table 1. Each drive route was about 5-8 minutes in duration at a speed of approximately 15 to 20 miles per hour. A 1 Watt PCS transmitter powered off the vehicle battery was connected to a 2.5 dBd magnetic mount antenna for the vertical mobile tests and to a "rubber duck" antenna fixed at approximately 45 degrees to simulate the operation of a portable unit. Each route was driven once for each antenna configuration (mobile and portable).

In the experiments, the receive antenna array consisted of purpose built hybrid antennas with 14 dB vertical, +45, and -45 degree antenna tilts. This configuration allowed simultaneous measurement of both the horizontal space diversity and the polarization diversity. Two antennas were spaced 10 feet apart to provide approximately 20-wavelength separation at 1800 MHz.

4. DATA ANALYSIS

Data were broken into 2000 sample (1-second blocks). Voltage was converted to power in dBm using a calibration table and then into power in Watts. A power average for each block of 2000 samples was computed.

Each block was sorted and ordered by increasing signal strength. The signal level at the 3% cumulative probability level (97% reliability level) was computed. This level is compared to the average for each record to provide an indication of the fading encountered in the record. For Rayleigh fading, the 3% cumulative probability level is approximately 15.8 dB below the average and is a good indicator of Rayleigh fading. This is illustrated in Figure 1. In the current round of experiments, maximal ratio combining of the diversity signals is simulated and the average and 3% cumulative probability level is computed for the combined signal, in other experiments, selection diversity gain was used as the metric.

Table 1. Description of Drive Routes

Route	Morphology	Distance (mi.)	Boresight angle
One	dense residential	0.5	45
Two	Residential	1	45
Three	Commercial	1.75	0
Four a	Urban	0.75	0
Four b	Urban	0.75	0
Five	Residential	1.5	45

The cross correlation index between the two vertical polarized signals, the two polarization diversity signals and between the cross-polarized signals and the vertical signals were computed using the technique described in Turkmani [1].

Over the entire route, the median of all the block averages is computed and compared for the combined and uncombined levels. Diversity gain is defined as the difference at the normalized 3% cumulative probability level between the combined signal and a reference signal which is defined as the 3% level of the cross polarized antennas or one of the vertical antennas. Figure 1. Illustrates the diversity gain effect. The first curve represents the cumulative distribution function of a single Rayleigh fading signal. The second curve represents the cumulative distribution of a signal with maximal ratio combining. The difference at the 3% cumulative level is the diversity gain. For Rayleigh fading, the theoretical diversity gain at the 3% level should be approximately 9.5 dB.

Figures 2 and 3 plot the difference between polarization antenna One and polarization antenna Two and vertical polarized antenna One and vertical polarization antenna Two in one second averages for drive route 2, which is a typical drive route.

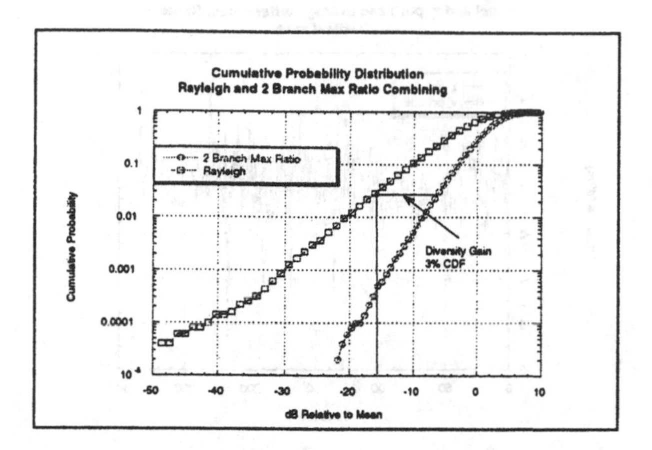

Figure 1. Cumulative Distribution for Rayleigh and 2 branch maximal Ratio Combining

This set of figures was indicative of what was observed in general. It was observed that the difference function for the polarization diversity antennas showed a slightly larger variation than did the horizontal space antennas. The polarization diversity system experienced a slightly higher correlation for the vertical mobile and a lower correlation for the random portable, which is consistent with the overall results.

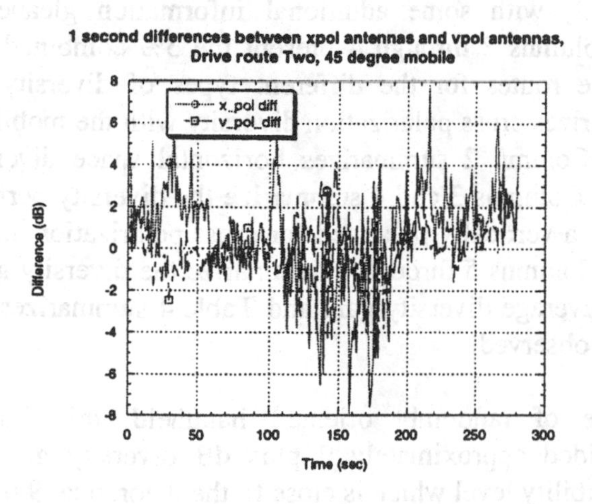

Figure 2

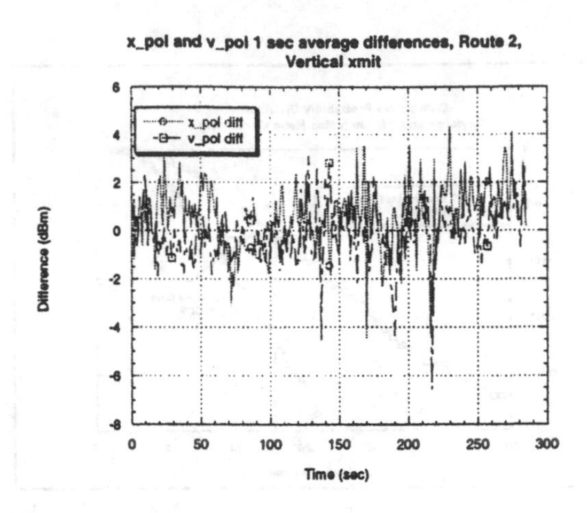

Figure 3.

5. DISCUSSION

Overall, the results of the experiment are summarized in Table 2 and indicate approximately the same results observed by Turkmani et al [1] and Weitzen et al [2] with some additional information gleaned from the experiments. Columns 2 through 4 present the 3% combined signal level over each of the routes for the different types of diversity combining. Column1 summarizes cross polarization diversity with the mobile antenna at random angles. Column 2 summarizes horizontal space diversity for the same conditions. Columns 3 and 4 summarize the diversity performance for the mobile antenna vertically polarized for cross polarization and horizontal space diversity. Columns 5 through 8 summarize the diversity gain. Table 3 summarizes the average diversity gain and Table 4 summarizes the average cross correlation observed.

For the case of randomly oriented handheld unit, both diversity techniques provided approximately 9 plus dB diversity gain at the 3% cumulative probability level which is close to the theoretical 9.6 for maximal ratio combining on a Rayleigh channel. In terms of bottom line performance, for the same case, the polarization diversity system performs approximately the same (less than 1 dB difference) as the horizontal space diversity system.

In the aggregate, for the case of the randomly oriented handheld unit, the cross polarization diversity system performs nearly the same as the horizontal space diversity system.

Table 2. Summary of 3% CDF Signal Thresholds and Polarization Gain

Route	Xpol 45	Space 45	Xpol mobile	Space mobile	Gain xpol 45	Gain space 45	Gain xpol 90	Gain space 90
One	-73.53	-72.84	-71.30	-67.60	8.42	8.21	6.92	9.02
Two	-83.14	-82.38	-82.43	-79.50	10.68	8.25	8.98	8.75
Three	-95.24	-95.15	-94.15	.91.96	9.14	8.52	7.48	7.92
Four a	-86.13	-85.00	-81.51	-77.40	9.62	8.79	7.78	8.87
Four b	-79.43	-78.12	-74.24	-74.81	7.51	7.79	7.10	9.01
Five	-91.97	-91.92	-89.74	-88.01	8.65	9.92	8.88	8.72

Table 3. Average Diversity Gains for Horizontal Space and Polarization Diversity using Portable and Mobile antennas

Avg Xpol portable gain	Avg Space Portable gain	Avg xpol mobile gain	Avg space portable gain
9.07	8.31	7.65	8.71

Table 4. Difference between Polarization Diversity and Horizontal Space Diversity and average diversity bottom line performance

Route	Xpol-Space portable	Xpol-Space mobile
One	-0.69	-3.60
Two	-0.76	-2.93
Three	-0.09	-2.19
Four a	-1.13	-4.11
Four b	-1.31	-3.43
Five	-0.05	-2.99
Average	**-0.67**	**-2.99**

Table 5. Cross Correlation Indices

Route	Xpol-port	Space-port	Xpol-mob	Xpol-port
One	0.24	0.26	0.5	0.16
Two	0.10	0.31	0.37	0.32
Three	0.14	0.30	0.27	0.25
Four a	0.12	0.08	0.44	0.15
Four b	0.09	0.15	0.47	0.14
Five	0.01	0.09	0.19	0.13
Average	**0.12**	**0.20**	**0.37**	**0.19**

For the vertical polarized mobile antenna, the vertical polarized, horizontal space diversity system out performs the polarization diversity system by approximately 3 dB in terms of bottom line performance and about 1.5 to 2 dB in terms of diversity gain. This is explained by polarization coupling which is 3 dB less into each of the 45 degree oriented antennas and some coupling between the vertical components in each of the antennas. Intuitively this makes sense since for the vertical mobile unit, receive antennas which best match the transmitted polarization provides the best results. For the randomly oriented handheld portable, any polarization slant at the base station is effectively the same.

The statistical summary of the experiment can interpreted as follows. For a mobile with random antenna orientation, the polarization diversity system performs approximately the same (less than a dB worse) as the conventional space diversity system. For the vertical polarized roof mounted mobile, the performance of the horizontal space diversity system with vertical polarization is approximately 3 dB better than the cross polarization system. In addition, for a vertically polarized signal, the median signal on either of the cross polarization branches should be 3 dB down from the vertical branch due to polarization coupling loss. The situation, in which there is a mix of vertically mounted vehicular mobiles and randomly mounted portables, occurs in cellular systems but not in PCS systems. In the cellular application, cross polarization diversity may be harder to justify unless the higher gain and ERP of the vehicle-mounted antenna offsets the loss in diversity gain.

It was also observed that the variance of the difference between the one-second averages in the polarization diversity branches was significantly greater than vertical branches.

6. POLARIZATION DISCRIMINATION

It has been demonstrated that polarization diversity reception can provide nearly identical performance as vertical polarization with space diversity for a wide class of portable subscribers. One of the questions currently under debate is what configuration should be used for the transmit antenna. The question is whether to use a separate vertical transmit antenna or can we transmit on one of the two polarization diversity antennas if the target receiver is a handheld portable. It is clear from the analysis, that if the target is vertical mobile, then we should transmit vertical to achieve optimal performance. In fact, if the target terminals are vertical mobiles, then polarization diversity should probably not be used.

The issue is whether and under what conditions there is significant discrimination, that is one polarization branch is favored over the other, which is bad, and under what conditions significant depolarization exists. To address this issue we selected several very short lateral drive routes, some of which were shadowed and some, which had a clear line of sight to the base station. A "rubber duck" antenna was mounted at 45 degrees and drives were made with the vehicle travelling in one direction, and then in the reverse direction to simulate changing the polarization sense.

Figures 4 and 5 plot vertical, cross polarization 1 and cross polarization 2, and vertical antenna 1, one second RSL averages for a drive approximately 100 meters from the antenna with a clear line of sight to the transmitter. The vehicle traveled at approximately 15 miles per hour. In the first figure we see that cross-polarized antenna 2 is favored and in the return drive cross polarization antenna 1 is favored. The vertical signal level was consistently between the two. The magnitude of the difference is on the order of 5 to 10 dB and indicates clear discrimination. When the range has increased to about 500 meters and there is no clear line of sight to the base station we observed that there is no clear polarization discrimination.

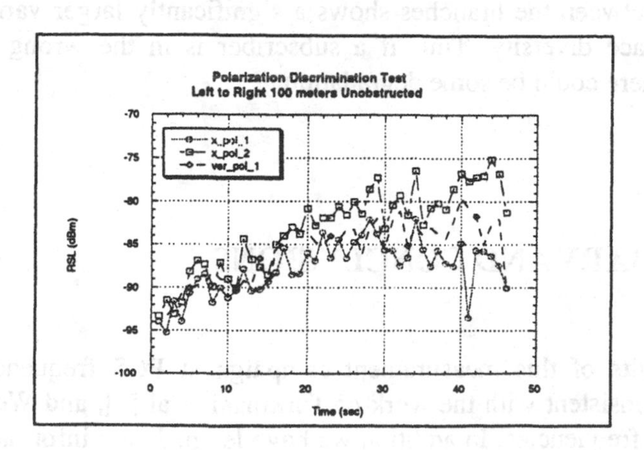

Figure 4 Polarization discrimination on a bridge

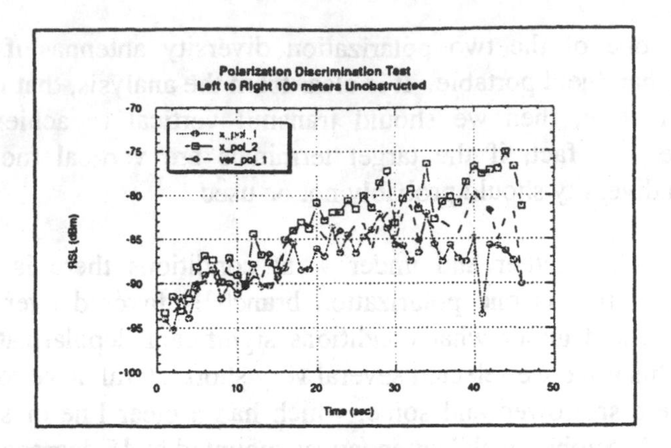

Figure 5. [Other Direction]

This trend was confirmed in the set of experiments and we can glean from the data the following observations. If we are at close range and there is a clear line of sight, there will be polarization discrimination that appears to be on the order of 3-10 dB. When in a scattering normal mode with no clear line of sight and at longer ranges, there appears to be a high level of depolarization, at most a couple of dB on the average, and the results indicate that either polarization sense would be adequate.

In a second set of experiments, we performed the parallel drives. In addition we stopped for 1 minute at the beginning and end point of each run. This allows static tests to observe what might happen in a wireless local loop situation. This exercise along with previous tests indicates that the difference between the branches shows a significantly larger variation than does the space diversity. Thus if a subscriber is in the wrong place and stationary there could be some degradation.

7. SUMMARY AND CONCLUSIONS

The results of this measurement campaign at PCS frequencies (1900 MHz) are consistent with the work of Turkmani et al [1], and Weitzen et al [2] at lower frequencies. In addition we have learned new information about the polarization discrimination phenomena and its effect on polarization diversity antenna systems.

- For a randomly oriented handheld unit, 45 degree cross polarization diversity reception provides on the average approximately the same performance as vertical polarization with horizontal space diversity to within about 1 dB. Polarization diversity has a number of financial and operational advantages that make it attractive in a number of deployment scenarios.

- For a vertically polarized mobile antenna such as vehicle mounted and WLL applications, polarization diversity is approximately 3 dB worse in overall performance than horizontal space diversity. Key in the application of cross polarization diversity to cellular systems, where there are a mix of portables and vehicle mounted mobiles, is whether the ERP advantage which a vehicle mounted unit has offsets the loss in diversity gain relative to a handheld unit. This also applies to a mixed system such as fixed Wireless Local Loop (WLL) and handheld units.

- Cross polarization discrimination appears to be a factor when there is a clear or near-clear line of sight to the base station and little scattering to cause depolarization. Overall there appears to be significantly greater variability in the variance of the difference between the branches for polarization diversity relative to space diversity. Cross polarization diversity may not be as effective for fixed WLL applications as space diversity.

- Using one of the cross polarization branches for transmit will perform adequately on the average, but there may be locations where there is significant degradation for a user whose antenna is oriented in the wrong direction. The occasional degradation must be weighed against the disadvantages of the vertical antenna configuration with horizontal space diversity. If a separate transmit antenna is required, then it should be vertically polarized.

- If the target receiver is the classic vehicle mounted vertical mobile, then vertical transmit and receive antennas should be used.

- H-V polarization diversity does not appear to provide good performance with a vertical mobile for either application.

Recommended Base Station Antenna Configurations:

Based on the analysis of data, we recommend the following standard configurations:

- If the target customers are predominantly vertical polarized mobiles, than use vertical polarized receive antennas with horizontal space diversity. If there is a duplexer, use one antenna for transmit and both for receive. If there is no duplexer then use standard cellular 3-antenna model.

- If the target customers use randomly oriented handheld portables or a mix of mobiles and portables, then we have several choices
  - Use 45-degree polarization diversity for receive, and a separate vertical antenna for transmit if a separate transmit is required.
  - Use 45-degree polarization diversity on receive and transmit on one branch when it is not convenient to use horizontal space diversity. Performance will be comparable, but there may be some slight degradation close to the site.

- If we are trying to mount on a monopole, then use a single unit configuration. Otherwise if mounting on a building, use individually mounted antennas.

- For most applications, use polarization diversity when it can save money or time in the deployment, as there is little difference between polarization and horizontal space diversity.

Acknowledgement

Measurement programs depend on the skill and diligence of the engineers and technicians who execute them. The authors wish to thank Tim Tatlock and Joe Krasucki for their dedication and attention to details that resulted in a high quality data.

REFERENCES

[1] A.M.D. Turkmani, A.A Arowojolu, P.A. Jefford, and C.J. Kellet, "An experimental Evaluation of the performance of two branch space and polarization diversity schemes at 1800 MHz", IEEE Transactions on Vehicular Technology, Volume 44, Number 2, May 1995

[2] J.A. Weitzen, J. W. Ketchum, J. Musser, "Comparison of Polarization and Space Diversity Antenna Systems in an Operational AMPS System", Proceedings, 2nd Conference on R&D in Massachusetts, March 1996

[3] R.G. Vaughn, and J. B. Andersen, "Antenna diversity in mobile communications", IEEE Trans. Vehicle. Technology Vol. VT-36, no 4, pp. 149-170, Nov. 1987

[4] W. C. Y. Lee, and Y.S. Yeh, "Polarization diversity system for mobile radio" IEEE Trans. Communication., Vol. COM-20, no. 5, pp. 912-923, Oct. 1972

[5] D.C. Cox, "Antenna diversity performance in mitigating the effects of portable radiotelephone orientation and multipath propagation", IEEE Trans. Communication, vol. COM-31, no. 5, pp. 620-628, May 1983

[6] J.D. Parsons, "The Mobile Radio Propagation Channel", Pentech Press, 1992, pages 140-145.

Chapter 3

USE OF SMART ANTENNAS TO INCREASE CAPACITY IN CELLULAR & PCS NETWORKS

MICHAEL A. ZHAO, YONGHAI GU, SCOT D. GORDON, MARTIN J. FEUERSTEIN

Metawave Communications Corp.

Abstract: The chapter examines three different smart antenna architectures and their real-world performance in cellular and PCS networks. The first smart antenna is designed for cdmaOne™ (EIA-95) CDMA cellular networks, with the goal of addressing many of the fundamental performance limitations that exist within these networks. The smart antenna is implemented as a non-invasive add-on (i.e. an appliqué) to current cdmaOne™ base stations, and improves capacity in CDMA networks through traffic load balancing, handoff management and interference control. Capacity improvements of greater than 50% have been achieved with the CDMA appliqué smart antenna through static and dynamic sector beam forming. The second smart antenna is designed as an integral subsystem embedded within specially adapted cdmaOne™ and 3G cdma2000™ base stations. The embedded architecture increases CDMA air link capacity by 100% to 200% through beam processing for each traffic channel. The third smart antenna, designed for current GSM networks, is implemented as an appliqué to existing base stations, and increases GSM air link capacity by 50% to 120% through increasing traffic channel carrier-to-interference (C/I) ratios, enabling increased fractional loading in frequency hopped networks.

cdmaOne is a trademark of the CDMA Development Group, cdma2000 is a trademark of the Telecommunications Industry Association (TIA)

1. INTRODUCTION

The wireless industry has witnessed explosive growth in subscribers in recent years. With the addition of wireless data services, such as email access and Internet browsing, many industry analysts predict that pressure on network resources will continue to dramatically increase. In order to provide sufficient resources for such a large demand, smart antenna systems are being deployed in large-scale fashion throughout major metropolitan cellular markets across the globe. These systems include multi-beam, dynamic sectorization and adaptive beam forming. Multi-beam techniques, (also referred to as fixed- or switched-beam) have been shown through extensive analysis, simulation, experimentation and commercial deployment to provide substantial capacity improvements in FDMA, TDMA, and CDMA networks [1-3]. Sector beam forming has been demonstrated to provide substantial capacity improvements in CDMA networks through static or dynamic sectorization [3-6]. Adaptive beam forming for each traffic channel can also provide significant capacity increases in wireless networks [3,7].

This chapter presents three smart antenna architectures and studies their capacity improvements in cellular and PCS networks. We begin by presenting a CDMA smart antenna, which as an appliqué has demonstrated significant capacity improvement through sector beam forming. The subsequent section describes an embedded CDMA smart antenna that performs baseband beam forming and switching for each traffic channel, an architecture that is integrated into the transceiver processing in the base station. Lastly, we describe a multi-beam antenna for GSM networks that is implemented as an appliqué to increase fractional loading in frequency hopped applications.

2. CDMA APPLIQUÉ SMART ANTENNA

When the question of CDMA smart antennas arises, it is clear from the literature that embedded techniques lead to significant capacity improvements when the phased-array processing is tightly interfaced with, or embedded within, the cell site's baseband transceiver processing [3]. An embedded smart antenna architecture for cdmaOne™ (EIA-95) networks is presented in Section 3. However, the large deployed base of cdmaOne™ cell sites does not have embedded smart antenna capabilities. Therefore an alternative to the embedded smart antenna is presented in this section; the alternative architecture is implemented as a non-invasive add-on to address the large deployed base of cdmaOne™ cell sites.

2.1 Network Performance

Over the past several years, cellular service providers have discovered that deployment, optimization and maintenance of CDMA networks are radically different from their now-familiar FDMA experiences. With unity frequency reuse, the difficult task of FDMA frequency channel planning goes away, replaced by the CDMA equivalent of per-sector PN offset reuse planning. Another facet of universal frequency reuse is the fact that every sector of every cell is either a potential handoff candidate or a possible interferer. In CDMA, there is no frequency reuse distance to separate co-channel interferers from one another. Due to local propagation conditions, it's not uncommon for a sector to overshoot the desired coverage area by several tiers of cells. With CDMA technology comes soft handoff, which provides a high quality make-before-break transition; but on the down side excessive handoff extracts forward link penalties in terms of higher transmit power requirements, increased interference, reduced capacity and potential dropped calls.

The golden rules necessary to achieve maximum performance from a CDMA network all involve interference control in one aspect or another. Successful optimization of the network, particularly the forward link, is an iterative process of making tough interference tradeoffs. For reliable call originations, dominant servers must be present, because calls originate on the access channel in a one-way connection; the same dominant server requirement is true for reliable handoffs, since an excessive number of potential servers can cause interference leading to dropped calls. In hard handoff regions (CDMA f1-to-f2 or CDMA-to-analog), managing interference is once again the key to reliable performance.

For over a decade now, the wireless industry has hotly debated capacity questions about CDMA technology. In reality, the capacity of a CDMA network is an ever-changing quantity that varies based on local topography and geographical traffic distributions over time. Network capacity is a strong function of the interference, as measured in terms of frequency reuse efficiency (ratio of in-sector interference to total interference), which is determined largely by local path loss characteristics. Network capacity is affected by spatial traffic density distributions; often these distributions are highly non-uniform and time varying on differing scales (hourly, daily, seasonally, event driven).

The smart antenna appliqué architecture is designed to provide CDMA cellular service providers with flexible tuning options for controlling interference, creating dominant servers, managing handoff activity, and handling non-uniform and time-varying traffic distributions. It also provides the ability to decouple the analog and digital sector configurations.

With smart antennas, a single physical array antenna can be used to synthesize completely different sector configurations for the digital and analog services. As the following sections will illustrate, there are strong theoretical and practical reasons that optimum CDMA sector settings are much different from optimum analog configurations; for example it may be desirable to implement CDMA as a 6-sector configuration while maintaining an underlying analog 3-sector network. Smart antennas enable such flexibility in deployment and optimization, while sharing a common antenna array for both analog and digital services, or among multiple digital services (e.g. vehicular voice, high rate data service, wireless local loop and private networks).

In cellular systems where antennas are shared between analog and CDMA, service providers are forced into fixed grid patterns due to the underlying frequency reuse assignments of the analog network. Without a smart antenna system, azimuth pointing angles of the sectors are locked into a rigid hexagonal grid pattern which forces all alpha, beta and gamma sectors—both analog and CDMA—to be aligned across the network. However, since CDMA is based on unity frequency reuse, there is no need to maintain a rigid grid pointing pattern across the entire CDMA network.

2.2 Traffic Load Balancing

Statistics derived from commercial cellular and PCS networks consistently indicate that traffic loads are unevenly distributed across cells and sectors. In other words, it's quite common for a cell to have a single sector near the blocking point, while the cell's other two sectors are lightly loaded. Traffic data from a number of cellular and PCS markets show that on average the highest loaded sector has roughly 140% of the traffic it would carry if all sectors were evenly loaded. By contrast, the middle and lowest loaded sectors have 98% and 65% of the traffic relative to a uniformly loaded case. Even though some sectors in a network may be blocking, significant under-utilized capacity exists in other sectors. The objective of traffic load balancing is to shift excessive traffic load from heavily loaded sectors to under-utilized sectors. The result is a significant reduction in peak loading levels and, hence, an increase in carried traffic or network capacity.

At a coarse level, static sectorization parameters can be adjusted for load balancing based on average busy hour traffic distributions. For optimum control of peak loading levels in time-varying traffic conditions, dynamic adjustment of sector parameters can be used employed on real-time measurements of traffic and interference. Under dynamic control, network parameters (neighbor lists, search windows, etc.) must be adjusted to support the range of dynamic sectorization control.

Both network simulation and experimental field results confirm that traffic load balancing can reduce peak loading levels, and thus minimize air interface overload blocking. An example presented in [5] describes a traffic hotspot scenario in which 54% of subscribers achieved acceptable service prior to smart antennas, while 92% of the subscribers obtained good service with smart antennas because of the ability to change sector azimuth pointing angles and beamwidths. Extensive commercial deployment results show an average 35% reduction in sector peak loading with sector beam forming, combined with additional benefits from handoff management and interference control.

2.3 Handoff Management

Cellular service providers often have an extremely difficult time controlling handoff activity. In CDMA networks, some level of handoff is desirable due to gains associated with the soft handoff feature (soft handoff allows the subscriber units to be simultaneously connected to multiple sectors). However, too much handoff can extract a significant performance penalty from the network. The penalty includes an increase in the total average transmit power per subscriber, which wastes valuable linear power amplifier (LPA) resources at the cell site, increases forward link interference levels and decreases forward link capacity accordingly. Excessive handoff activity can also result in dropped calls due to handoff failures.

For optimum forward link capacity, CDMA network operators strive to tightly manage the amount of handoff activity. Typical networks may run at handoff overhead levels between 65% and 100% (i.e., 1.65 to 2.0 average handoff links per subscriber. Smart antennas can be used to manage handoff activity by controlling the RF coverage footprint of the cell site to tailor handoff boundaries between sectors and cells, and reducing rolloff of sector antenna pattern. Figure 1 (left) illustrates the radiation pattern from a smart antenna versus an off-the-shelf commercial sector antenna. With smart antenna arrays, it is possible to synthesize radiation patterns with sharp rolloff (i.e. steep transition out of the antenna's main lobe) in order to reduce handoff overhead, while still maintaining coverage. Sector patterns synthesized from the phased array antenna can be much closer to an ideal sector pie-slice or conical pattern. Commercial deployment results demonstrate that smart antennas can reduce handoff overhead by 5 to 15%, thus increasing forward link capacity by an equivalent amount.

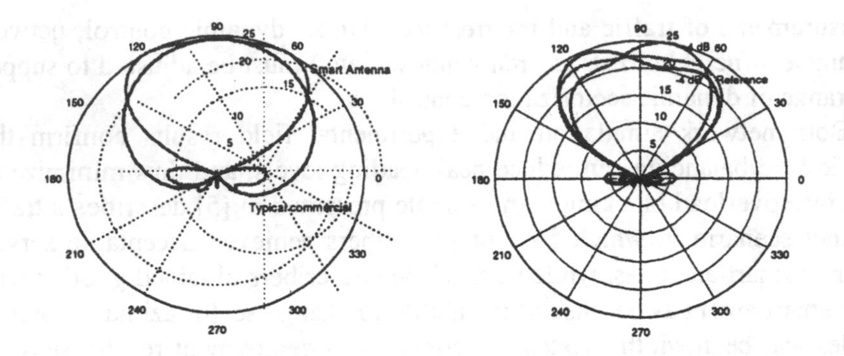

Figure 1. Left: Radiation pattern of synthesized sector pattern from smart antenna versus typical commercial off-the-shelf antenna. Right: Synthesized sector antenna patterns from smart antenna illustrating sector sculpting to control interference.

2.4 Interference Control

As mentioned previously, the most fundamental aspect associated with tuning CDMA networks is managing interference levels. On both the forward and reverse links, varying interference levels across the network mean that coverage, quality and capacity change based on local geography and time-of-day. The best way to illustrate the sensitivity of reverse link capacity to antenna characteristics is through the reverse link frequency reuse efficiency (ratio of in-sector to total interference). Reverse link capacity is directly proportional to the frequency reuse efficiency. A simulation presented in [5] shows the following results for a hexagonal grid cell layout with the Hata propagation model. At wide antenna beamwidths, reuse efficiency is low due to the large sector aperture resulting in the capture of significant interference from subscribers in other sectors and cells. At narrow beamwidths, reuse efficiency is low due to reduction in main beam coverage area combined with the impact of antenna sidelobes. For the simulation example, beamwidths of roughly 70° to 90° result in the best reverse link capacities. Area coverage probabilities from the simulations show that the design target of 97% area coverage probability target is maintained over this range of antenna beamwidth as well.

During initial network installation and subsequent network maintenance, service providers spend a significant amount of time and effort to fine-tune interference levels. Operators may adjust transmit powers, downtilt antennas, change antenna patterns, or modify network parameters to eliminate interference from problem areas. Smart antennas provide an unprecedented degree of flexibility in tuning the RF coverage footprint of each sector. Figure 1 (right) illustrates several of the sector antenna patterns that can be created by sculpting the coverage with sector beam forming. In the figure, three radiation patterns are shown: the reference case is the unadjusted sector pattern; the other two patterns show +4 dB and –4 dB adjustments in particular azimuth directions. Transmit power can be increased in specific directions to enhance coverage in traffic hot spots and inside buildings, or to create dominant servers in multiple pilot regions. In other directions, transmit power can be reduced to minimize interference, control handoff activity or alleviate severe cases of coverage overshoot. Using smart antennas to control sector footprints is significantly more flexible than the alternatives of employing antenna downtilts or adjusting sector transmit powers—adjustments that impact the entire coverage area of the sector, rather than confining changes to the specific problem spot.

2.5 Commercial Deployment Results

Extensive commercial deployments of the CDMA appliqué allow the capacity improvements to be quantified. The capacity increases are due to three factors: traffic load balancing, handoff overhead reduction and interference control. Detailed capacity models and measurement techniques have been developed to estimate capacity improvements with and without smart antennas [8]. Figure 2 illustrates a real-world deployment example of capacity improvement using the smart antenna. The scatter plot shows CDMA forward overload blocking due to capacity exhaustion, versus carried traffic as measured in primary Walsh code usage (summed Erlangs). Measurements were made using mobile switching center (MSC) statistics collected from live commercial traffic. The baseline case (solid line, solid diamonds) is without smart antenna; the smart antenna case (dashed line, open circles) is with the smart antenna. As observed from the plot, the smart antenna allows significantly more traffic to be carried at lower levels of overload blocking, hence increasing the CDMA air link capacity. In this example the capacity improvement is over 30%, as can be observed since the smart antenna trend line is shifted down and to the right compared to the baseline case (meaning that more traffic is handled at lower blocking levels, hence higher capacity). The measured capacity increases are extremely close to those predicted with analytical models of CDMA link capacity [8].

Table 1 provides examples of typical commercial deployment results for the CDMA appliqué smart antennas at 10 different cell sites within 5 separate networks. Capacity increases of over 50% have been achieved. While demonstrating an average of 40% capacity increase, the smart antennas also simultaneously showed averages of 28% and 17% reductions in dropped calls and access failures. Capacity was increased while maintaining or improving quality and coverage.

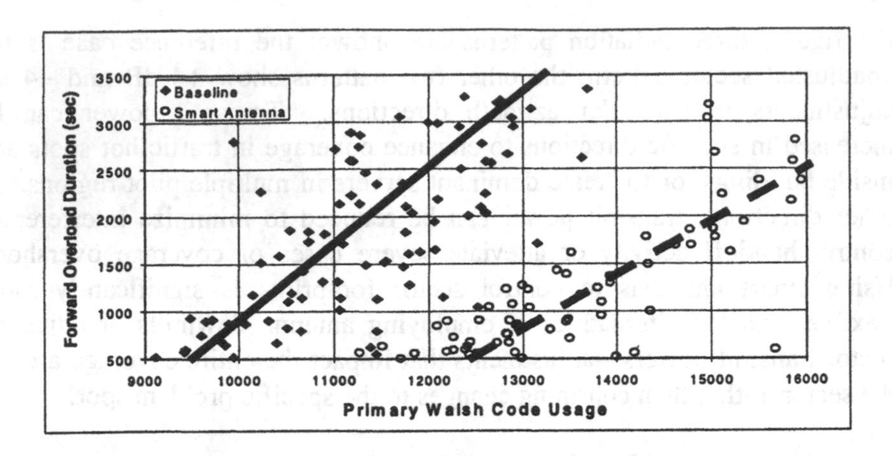

Figure 2. Scatter plot of forward overload blocking versus carried traffic.

Table 1. Commercial deployment results for CDMA appliqué.

Cell Site	Network	Measured Capacity Increase
1	A	51%
2	B	53%
3	B	45%
4	B	30%
5	C	46%
6	C	35%
7	C	32%
8	D	40%
9	D	35%
10	E	33%
	Average	40%

3. CDMA EMBEDDED SMART ANTENNA

The CDMA appliqué smart antenna examined in the previous section is designed for improving capacity of current cdmaOne™ base station transceiver systems (BTSs). For next generation cdmaOne™ and 3G cdma2000™ base stations, it is possible to increase capacity further by tightly coupling the smart antenna processing within the BTS's baseband transceiver processing [3,7]. Capacity is further increased by forming best transmit and receive patterns for each traffic channel, rather than for each sector as is done by the appliqué.

This section examines an embedded smart antenna architecture for cdmaOne™ and cdma2000™ networks. The architecture is designed as an extension of current base station designs, rather than a complete re-design of traditional CDMA BTSs. The embedded CDMA smart antenna re-uses as much of the existing component parts of a traditional BTS as possible, and supports two deployment configurations: traditional and smart antenna.

3.1 Traditional Base Station

It is helpful in understanding the embedded smart antenna to first examine the architecture of a traditional cdmaOne™ BTS. Figures 3 and 4 show the architecture and signal flow of a traditional BTS that consists of three parts: antenna interface, RF processing and baseband processing. The antenna interface comprises six traditional sector antennas. Three are used as both transmit and receive antennas in a duplexed configuration; the other three are only used as receive antennas for spatial diversity.

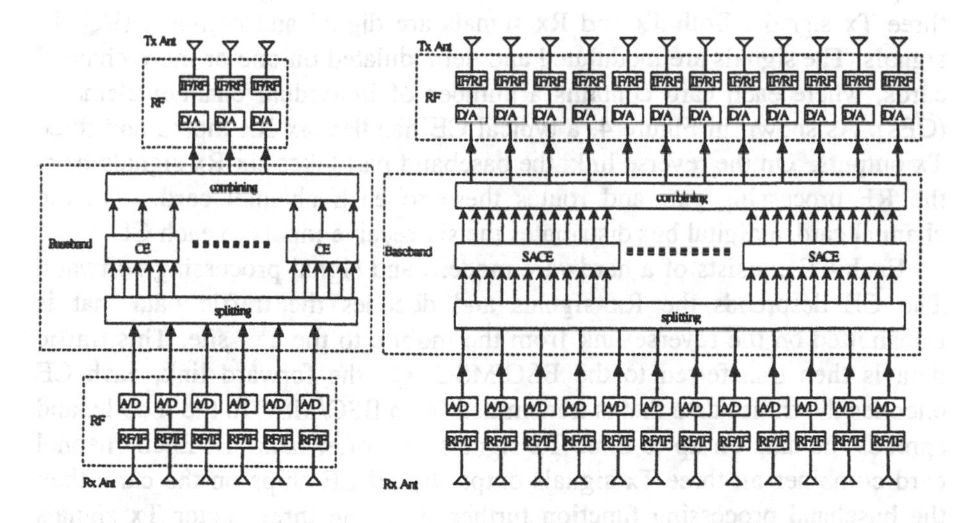

Figure 3. Signal flow of BTS (left) and embedded smart antenna BTS (right).

The RF processing section interfaces between the antennas and baseband processing part. In simplified form, each RF receive path consists of radio frequency to intermediate frequency (IF) conversion circuitry followed by analog to digital conversion. Each RF transmit path includes a digital to analog converter followed by intermediate frequency to radio frequency conversion circuitry.

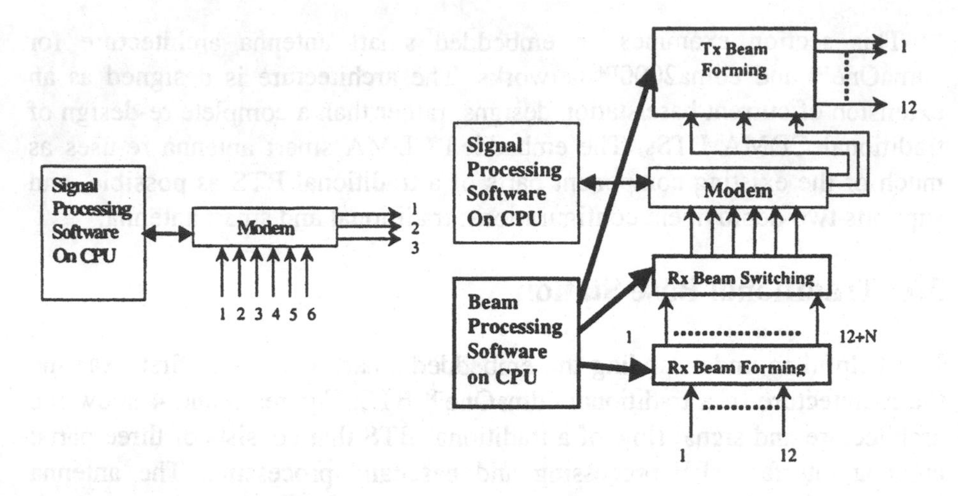

Figure 4. Standard channel element (left) and smart antenna channel element (right).

The baseband processing part interfaces between the RF processing and the base station controller (BSC) or mobile switching center (MSC). On the RF processing side, the baseband part receives six Rx signals, and transmits three Tx signals. Both Tx and Rx signals are digital and complex (I/Q) IF signals. The signals are modulated and demodulated on one or more channel cards, where each card contains a number of individual channel elements (CEs). As shown in Figure 4, a typical CE handles six Rx inputs and three Tx outputs. On the reverse link, the baseband part takes six Rx signals from the RF processing part and routes these to each channel card. On the channel card, a digital bus distributes the six receive inputs to each CE.

Each CE consists of a hardware modem and signal processing software. The CE despreads the Rx signals and decodes the traffic data that is transmitted on the reverse link from the mobile to the cell site. This traffic data is then transferred to the BSC/MSC. On the forward link, each CE encodes the traffic data that is transmitted from BSC/MSC to the mobile, and spreads the data on up to three Tx signals for softer handoff. Each channel card combines all three Tx signals output by all CE chips on the card, then the baseband processing function further sums the three sector Tx signals from multiple channel cards.

3.2 Embedded Smart Antenna Base Station

The right hand sides of Figures 3 and 4 show the architecture and signal flow of the embedded smart antenna in a cdmaOne™ BTS. As is evident from both figures, the addition of the embedded smart antenna is an incremental evolution of the cdmaOne™ BTS, making maximal reuse of existing components and subsystems. In Figure 3, the number of Rx paths has been increased from 6 to 12, to convert from a typical three sector configuration to support a 12 element array. The number of Tx paths has been increased from 3 to 12, again to support a 12 element array.

The Rx and Tx processing chains remain essentially unchanged, except that the total number of paths has been increased. Because of the array architecture with multiple antenna elements comprising a sector, the average power of each PA is approximately ¼ of the average power of a PA used in the traditional BTS. The smart antenna subsystem also includes gain and phase calibration circuitry for all 12 paths to support accurate weighting coefficients for beam forming.

The baseband processing part is an enhanced version of that found in a traditional BTS. Each channel card transmits and receives 12 signals. Adding smart antenna processing components augments each CE on the card. These smart antenna channel elements (SACEs) are described below. 12 Tx and Rx signals are routed through each SACE.

As shown in Figure 4, each SACE inputs 12 Rx signals and outputs 12 Tx signals. The SACE consists of a traditional channel element plus several new components, including the following: an Rx beam forming unit, an Rx beam switching unit, a Tx beam forming unit, and beam processing software. The Rx beam forming unit transforms the twelve Rx signals into at least twelve Rx signals, where each signal represents the output of a formed beam. The Rx beam switching unit switches six Rx signals to the modem.

For each Tx signal from the modem, the Tx beam forming unit transforms it into 12 components, which coherently form a Tx beam. For the three sector Tx signals from the modem, the Tx beam forming unit combines their transformed signal components into one output Tx signal. For example as shown in Figure 4, the Tx beam forming unit can create interstitial beams to overcome cusping (i.e. cross-over) loss associated with a traditional fixed-beam antenna. In addition the Tx beam former can be used to create custom radiation patterns for auxiliary pilots in more advanced cdma2000™ networks.

The transformation coefficients used by the Rx and Tx beam forming units in each SACE are initialized and periodically updated by the beam processing software. The beam processing software also controls the Rx beam switching. The software receives search data from the modem through the baseband signal processing software, and then determines the Rx beam switching decision and Tx beam transformation coefficients. More specifically, the software compares measurements of the despread signal-to-interference ratio from the six modem input signals. The software algorithm controls the Rx beam switching unit to continuously provide the best six of the available inputs as signals to drive the modem. The software also keeps track of which demodulator elements in the modem are actually assigned and locked on to receive paths to avoid switching any of the logical inputs that are actively being demodulated.

For each Tx signal from the modem, the software determines the optimum Tx beam transformation coefficients. If the Tx signal is a pilot, synchronization or paging channel to be broadcast over an entire sector, the Tx beam created is an antenna pattern for the sector. If the Tx signal is a traffic channel, the transformation coefficients are determined based on the historical characteristics of the Rx beam switching unit settings and Rx beam transformation coefficients, in order to minimize the interference to other mobiles.

3.3 Advantages of Embedded CDMA Smart Antenna

The primary advantage of the embedded CDMA smart antenna is the capacity increase derived from improving uplink and downlink C/I ratios. A number of different simulation, analytical and measurement models can be used to estimate the capacity improvements associated with smart antennas [3,7,9]. For a 12 beam configuration, capacity increases in the range of 100% to 200% can be expected for the embedded CDMA smart antenna [3,7]. As predicted, reverse link transmit power reductions of 3 to 4 dB have been measured during CDMA smart antenna field tests in dense urban environments [9].

Compared to a traditional cdmaOne™ or cdma2000™ BTS, the embedded smart antenna equipped BTS has the following additional advantages. The service provider has increased flexibility because the BTS can be deployed in a standard configuration or in a smart antenna configuration. Initially, if capacity is not constrained, the BTS can be deployed without the smart antenna; then when traffic builds, the BTS can be field upgraded to support the smart antenna. In addition, the embedded smart antenna supports traffic load balancing, handoff overhead reduction and interference control as described in Section 2 for the appliqué.

4. GSM APPLIQUÉ SMART ANTENNA

4.1 Network Performance

The maximum number of mobiles that a network can physically support defines the capacity, which may be limited by either the network hardware or RF air link. The GSM system has its own requirement on the minimum carrier-to-interference (C/I) and signal-to-noise ratio (SNR) to maintain a high quality communication link. An active call may drop when there are no channels available with C/I values greater than the minimum requirement. Unlike a CDMA network, however, a GSM mobile user only generates interference to those mobiles served by co-channel frequencies, but never to those in the same cell as shown in Figure 5. Given limited spectrum, as the number of mobile users on a BTS increases, more co-channel interference is generated. This increase in co-channel interference serves to limit the ultimate capacity of the network.

Each network operator has finite GSM spectrum. As traffic increases, eventually the network operator will run into capacity limitations. Increasing the number of radio transceivers (TRXs) in a network is the least expensive way to increase the capacity, if the co-channel interference limited capacity bound has not been reached. As the interference bound is exceeded, other techniques must be employed to increase capacity.

Consider the case shown in Figure 5, where Cell A and Cell B are co-channel with each other, assuming a frequency reuse of N=7 and omni-directional antennas. Cell A is in a traffic hot-spot and Cell B is in a lower traffic region, or a warm-spot. Further assume that the maximum number of TRXs that can be installed in a cell site is 6, due to spectrum limitations. In the Figure 5 example, only 5 TRXs at Cell A have been installed due to actual traffic loading in the hot-spot, and 4 TRXs at Cell B in the warm-spot. When Cell B runs into capacity limits, the least expensive way to exploit the capacity headroom is to add an extra TRX at Cell B. The level of co-channel interference at the hot-spot in Cell A immediately goes up due to the increased co-channel traffic at Cell B. Therefore, more soft-blocking (i.e. co-channel interference) happens in the hot-spot. Going further, one more TRX is added into Cell A to make the number of TRXs reach the maximum of 6. Although the extra TRX temporarily compensates for the interference induced capacity loss at Cell A, the increased traffic drives up the level of co-channel interference to Cell B. Eventually, when Cell B runs out of capacity again, a new round of adding an extra TRX in Cell B starts. Unfortunately, since Cell A already has the maximum number of TRXs, the increased level of interference in Cell A causes additional soft-blocking. In this situation, the service provider must use other solutions to increase the capacity at Cell A, since adding TRXs is no longer an option.

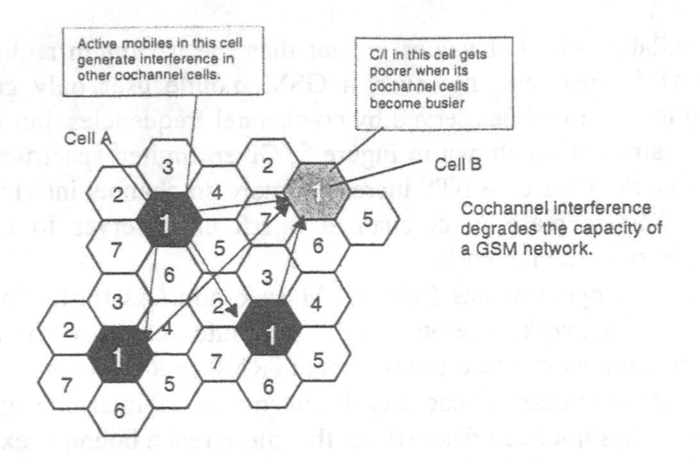

Figure 5. Co-channel interference limits the capacity of a GSM network (reuse N=7,Omni)

4.2 Network Optimization for Capacity

For a service provider to be successful in the wireless marketplace, the network must be continuously optimized to maintain sufficient capacity to support the growing customer base. Optimising capacity involves increasing hardware investment, as well as reducing the level of co-channel interference in the entire network. The GSM standard supports a number of methods for improving C/I ratios, such as:

- Discontinuous transmission (DTX) and dynamic power control (DPC) to minimize transmitted power.
- Frequency hopping to exploit frequency diversity by more uniformly distributing interference to co-channel cells on the downlink, and reducing slow fading impact on the uplink.
- Sectorization of cells to reduce the co-channel interference levels on the downlink, and improve the C/I ratio on the uplink at the expense of reduced trunking efficiency.
- Diversity reception on BTS uplink to improve signal quality and reduce the minimum C/I requirement.

Operators generally use these techniques along with other network planning and engineering tools, such as micro/picocells, overlay/underlay, fractional loading and frequency reuse planning, to turn improved C/I ratios into additional network capacity.

Most GSM networks in the world have been in use for more than 5 years. After so many years of optimization, most operators have gone through several cycles of network modifications, based on the techniques mentioned above. To increase the number of BTSs in a network the minimum distance between BTSs has been reduced to less than 300 m in many metropolitan areas. Traffic hot-spots in downtown areas and dense business districts have become the most problematic and troublesome regions. As an example, statistics of most large GSM networks show the following trend [10]: 1% of cells generate 10% of the total interference, 3% generate 20%, 8% generate 40%, and 13% generate 50%, respectively.

Fixing interference problems in these few hot-spot regions can significantly reduce the overall network interference level. Due to limited spectrum and co-channel interference limitations, increasing capacity in these hot-spot regions cannot generally be accomplished with traditional frequency planning tools and drive testing. In many circumstances, using smart antennas in part of the network is the most cost-effective way to resolve traffic hot-spot problems.

4.3 GSM Appliqué Smart Antenna

A GSM smart antenna system is specifically designed for the reduction of co-channel interference, using sophisticated DSP algorithms and array antenna technology. In general, smart antenna technologies can be classified in three categories: multi-beam, adaptive beam forming and co-channel interference cancellation. Multi-beam and adaptive methods have been presented in previous sections of this chapter. Interference cancellation exploits multi-user detection algorithms to improve C/I, often resulting in better C/I gain compared to the other two methods. Unfortunately, it is best applied to time division duplex (TDD) system, such as Japan's PHS (personal handyphone system). Multi-beam systems switch among a set of fixed, narrow beams. Adaptive beam forming uses a DSP engine and an accurately calibrated RF front-end to create optimum weighting coefficients. Although adaptive beam forming may produce better performance than a multi-beam system, multi-beam has proven to be more feasible and cost-effective for FDD air interfaces than purely adaptive approaches. Multi-beam techniques offer the following advantages over adaptive beam forming and interference cancellation:

- Can be implemented as appliqué to an existing BTS because it is transparent to received/transmitted RF signals.
- Economical tradeoff of complexity of RF hardware and DSP engine.
- Compatible with packet data standards, such as GPRS (general packet radio services) and EDGE (enhanced data rates for global evolution).

Figure 6 shows a block diagram of the GSM smart antenna using multi-beam technology. The fundamental components of the system are the narrow beam antenna array and the RF beam switching matrices.

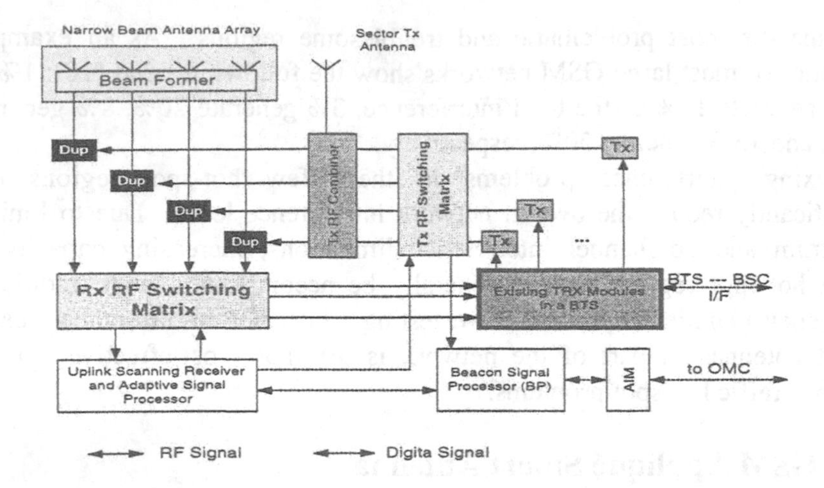

Figure 6. GSM appliqué smart antenna architecture.

The narrow beam antenna array generates 4 beams to cover a traditional sector, where each beam covers ¼ of the sector. A customized Butler matrix is used in the array as a beam former. Several models of the antenna are available to match different sector widths, for example in a 90° tri-sector BTS, each beam covers 22.5°. To maintain the sector beacon carrier coverage, the system uses an independent, but integrated, sector antenna for broadcasting. The smart antenna uses a beacon processor to derive system information and timing from the BTS. The uplink scan receiver analyzes the uplink RF signal for each call, using DSP algorithms to select the best antenna beams. Best antenna beam decisions are then relayed to Rx and Tx RF switch matrices. The smart antenna supports the advanced features defined by the GSM standard (DTX, DPC, baseband and synthesized frequency hopping) as well as a GPRS pass-through capability.

The GSM smart antenna uses high-power switching on the downlink to allow re-use of the existing single-carrier power amplifiers in the cell site. The smart antenna interfaces to the cell site using standard RF cable connections to the TRX modules, and includes remotely accessible operations, administration and maintenance functions.

4.4 Field Trial Results

A field trial of the GSM appliqué smart antenna was conducted in a dense urban portion of a large GSM network. The field trial system used three narrow beams to cover a traditional 90° sector. Installation of the smart antenna was transparent to the cell site, requiring no changes at the BTS or network switching elements. A reference sector antenna was used to measure the C/I improvement of the smart antenna, with an observed average uplink improvement of 6 to 9 dB. During the smart antenna trial, the MSC statistics recorded a dropped call rate reduction of over 50%, compared to the period before the smart antenna was installed.

5. CONCLUSIONS

This chapter has presented techniques for using smart antennas to increase capacity in both CDMA and GSM networks. Three architectures were presented: a CDMA appliqué smart antenna, a CDMA embedded smart antenna, and a GSM appliqué smart antenna. Commercial deployments and field trial results have demonstrated the ability of smart antennas to significantly increase the capacity of real-world cellular and PCS networks.

The CDMA appliqué smart antenna was presented as a non-invasive add-on to current cdmaOne™ base stations. The design is non-traditional in the sense that it does not attempt to create an optimum antenna pattern for each traffic channel, but rather beam form on a per-sector basis. Each sector is assigned radiation patterns exhibiting optimized beamwidths, azimuth pointing angles, and sculpting characteristics, with the objectives of balancing traffic load, managing handoff activity and controlling interference. Capacity increases of over 50% have been demonstrated in real-world commercial deployments.

The CDMA embedded smart antenna was presented as an integral subsystem within next generation cdmaOne™ and cdma2000™ base stations. The embedded smart antenna extends the sector beam forming technique to provide beam forming for each traffic channel. The smart antenna can increase the CDMA air link capacity by 100% to 200%, through increasing traffic channel carrier-to-interference (C/I) ratios.

The GSM appliqué smart antenna can significantly improve the C/I performance of an existing network. Simulation results show that capacity growth of 50% to 120% can be achieved with less than 38% penetration of smart antennas within the network. A field trial has demonstrated C/I improvement of 6 to 9 dB using the narrow beam switching technology. More than 50% dropped call rate reduction was reported by MSC statistics during the trial. Ericsson has reported similar conclusions, where a field trial demonstrated a 25% network capacity improvement with a smart antenna penetration of 5% in the network [11].

REFERENCES

[1] Y. Li, M. J. Feuerstein, D. O. Reudink, "Performance Evaluation of a Cellular Base Station Multi-beam Antenna", *IEEE Trans. Veh. Tech.*, Vol. 46(1), Feb. 1997.

[2] M. J. Ho, G. L. Stuber, M. D. Austin, "Performance of Switched-Beam Smart Antennas for Cellular Radio Systems", *IEEE Trans. Veh. Tech.*, Vol. 47(1), Feb. 1997.

[3] J. C. Liberti, T. S. Rappaport, *Smart Antennas for Wireless Communications: IS-95 and Third Generation CDMA Applications*, Prentice Hall, 1999.

[4] T. W. Wong and V. K. Prabhu, "Optimum Sectorization for CDMA 1900 Base Stations", *Proc. IEEE VTC'97*, May 1997, Phoenix, AZ.

[5] M. J. Feuerstein, J. T. Elson, M. A. Zhao, S. D. Gordon, "CDMA Smart Antenna Performance", *1998 Virginia Tech Wireless Symp.*, June 1998, Blacksburg, VA.

[6] M. Mahmoudi, E. S. Sousa, H. Alavi, "Adaptive Sector Size Control in a CDMA System Using Butler Matrix", *IEEE VTC'99*, May 1999, Houston, TX.

[7] M. Zhao, M. Feuerstein, "Exploiting Smart Antennas to Increase Capacity", *3rd Int. Summit China '99.*, Nov. 1999, Beijing, China.

[8] S. D. Gordon, M. J. Feuerstein, M. A. Zhao, "Methods for Measuring and Optimising Capacity in CDMA Networks Using Smart Antennas", *1999 Virginia Tech Wireless Symp.*, June 1999, Blacksburg, VA.

[9] M. J. Feuerstein, D. O. Reudink, "Multi-beam Smart Antenna System Performance in CDMA Networks", *CDMA Tech. Conf.*, Sept. 1997, Dallas, TX.

[10] H. Goodhead, "Breathing New Life into Ageing Networks—a Review of Traditional and New Techniques in Planning and Optimization", *Proc. of IBC Conf. on Optimizing & Upgrading BSS*, December 1999, London, U.K.

[11] H. Dam et al, "Performance Evaluation of Adaptive Antenna Base Stations in a Commercial GSM Network", *Proc. IEEE VTC'99*, May 1999, Houston, TX.

PART II

DEPLOYMENT OF CDMA BASED NETWORKS

Chapter 4

OPTIMIZATION OF DUAL MODE CDMA/AMPS NETWORKS

VINCENT O'BYRNE*, HARIS STELLAKIS**, RAJAMANI GANESH**

**GTE Service Corporation,. 40 Sylvan Rd., Waltham MA 02451-1128,*

***GTE Laboratories, 40 Sylvan Rd., Waltham MA 02451-1128*

Abstract: The optimization of CDMA deployed over the legacy analog network is shown to be a complex series of issues and a function of both the intrinsic capacity of CDMA and some practical considerations associated with the incumbent AMPS system. The implications of voice rate, voice quality, deployment strategy and timing on overall network capacity are analyzed together with practical issues of managing the interference between the two networks with use of appropriate guard zones and guard bands etc. The maintenance of call quality, handdown and flawless network operation in the border areas of the dual-mode and analog network are analyzed by including appropriate usage of border and beacon cell sites and CDMA equipment parameters such as cell size. Some of the options available to the network provider in order to balance and optimize the trade-offs between coverage, quality and capacity are analyzed together with how best to proceed in an on-going basis as the growth of both analog and CDMA traffic continues.

1. INTRODUCTION

The underlying system of cellular carriers in the late eighties and early nineties was based on analog technology. However during the early part of the past decade, rapid growth of wireless user traffic has made stringent demands on capacity for all wireless systems throughout the US. The cellular service providers realized that the analog system AMPS , which has been in service in North America since the late 1970's, was no longer able to provide the required capacity and quality economically. The AMPS system was originally deployed on a regular hexagonal arrangement of cells [1] which allowed frequencies to be reused after some "re-use" distance defined by consideration of the desired signal quality and the propagation environment. Early deployments tended to mimic loosely the theoretical arrangement of sectors with the spectrum and channels split up into recognizable groups based on a 7/21 reuse factor [2]. These network wide assumptions of the propagation environment and channel groupings resulted in inefficiencies being built into the analog system which would later be unleashed, by employing individual frequency planning techniques to help clear spectrum for the deployment of digital technologies. However at the time they were expedient measures to get up and running as soon as possible.

The explosive growth in demand led to increased burden on the analog sites and also increased blocking for the subscribers. The AMPS system started to become capacity limited and the cell site RF infrastructure was tuned to limit the interference to other cell sites. The issue of how to address this need for additional capacity was in the form of short-term solutions like cell splitting, sectorization and in the case of the longer term the deployment of a newer technology with higher capacity. However in many cases because of user mobility these newly created sites started to block as soon as they were deployed.

The shorter-term solution had some basic problems as it was getting harder in some markets to deploy more cell sites because of zoning issues and typically the planing-to-deployment phase of new cells was relatively long. Also, a large majority of cell sites, especially in urban areas had been sectorized already. The increase in the number of cells in the field also brought with it frequency planning issues which had to be resolved. These solutions had been the only option for a long time but it was becoming economically prohibitive to continue down this path.

A longer-term solution had to be found. This was in the form of newer technology, namely digital. The option on the table at the time was Time Division Multiple Access (TDMA), standardized as IS-54, which had been quoted as having upto 3[1] times the capacity of an analog system. This was seen by many vendors as the means to increase their capacity and alleviate the pressure on their main system. TDMA was also seen as an easier migration strategy with the ability to utilize the frequency planning technology that was already present in the analog networks.

Another technology that was being considered was Code Division Multiple Access (CDMA) which came upon the scene as TDMA was being standardized. Though this technology was in use by the military since the early 1920s, it was commercialized by Qualcomm and later standardized by TIA/EIA for the digital cellular (IS-95 [3]) frequencies at 800 MHz and PCS bands at 1900 MHz. Apart from offering solutions to the problems of frequency reuse and interference rejection, spread spectrum CDMA also provides benefits such as multipath immunity and low probability of intercept (LPI). Initially, the CDMA system based on the IS-95 standard was thought to be able to provide 10 to 20 [4,5] times the capacity of analog technology[2], with the former being the more accepted number at the time. However, deployments until the present day have shown that the actual capacity gains are much lower than those expected initially – closer to 6 or 7 times that of the AMPS system. These reduced gains are more a consequence of the manner in which it has been deployed as opposed to any technical or ideal obtained value. It also had fewer vendor choices, especially in 1993, when the only vendors were Qualcomm, Lucent and Motorola. This technology though behind the curve from a standardization point of view, did offer the prospect of a longer-term solution. The capacity advantage, together with other features such as the possibility of "overlay" operation, improved voice quality, make-before-break connections with soft and softer handoff capability and the ability to take advantage of multipath was seen as a necessity by many operators.

When CDMA was put forward as the answer to the capacity problems of the analog system at 800 MHz, it was envisioned that CDMA would be deployed in an optimum fashion. However, in many cases the CDMA system was deployed as an overlay to the present analog system, using the same cables and antennas. Over the years the analog system had been tuned and retuned and in general, especially in high traffic areas, was designed to be capacity limited. Antennas, amplifiers etc were tilted or tuned down to

[1] This does not include the overhead channels.

[2] Assuming the lower rate vocoder at 9.6kbps

reduce interference with neighboring sites. CDMA, because of its inherent greater capacity would be deployed in a more coverage-limited manner. Historically also there was little consideration given to the overlap between the analog sectors of the same site, due to the relatively large frequency reuse of analog. However, for CDMA systems, this overlap results in a reduction in the sectorized gain of CDMA [7] and increased use of cellular channel resources. This reduced the deployment efficiency. It could also not take full advantage of its improved link budget of between 3 and 5 dB as deployments required that for a hand-down to occur seamlessly that the CDMA and analog sites be collocated and have effectively the same footprint. These have resulted in realizable gains of between 6 and 10 [5,6] when deployed over an analog network, depending on the data rate set employed.

At this time both TDMA and CDMA used an 8 kpbs vocoder, which was judged by their subscribers to be inferior to analog but offered the greatest capacity increases at "acceptable" voice quality levels, to the service provider. Ameritech [8] conducted joint trials with both TDMA and CDMA in its St. Louis market in 1992 and found that customers were not happy with the 8kbps TDMA voice quality, and that CDMA voice quality was preferable. However, not all cellular markets or operators were equally congested and those operators in less dense markets felt it prudent to wait for CDMA standardization and even for a better vocoder (at the time work was starting on a higher "Toll Quality" vocoder operating @ 13 kpbs). By so doing they hoped that they would make the migration to digital technology only once as many felt that TDMA was only an interim solution and that ultimately CDMA was the best long term solution. At around this time the industry had become much more competitive and the need for a differentiator was becoming obvious.

From a subscriber's point of view it was also becoming necessary and desirable to be able to roam freely across the country. The only common standard between the carriers in the cellular bands was analog. So analog became the means to allow roaming between operators who had deployed different digital technologies. Given this need to maintain the analog network the options available to, and benefits obtainable by the provider are dependent on the deployment that has taken place to-date.

As an example, Figure 1 [ref. 9] illustrates the number of sites required as a function of various deployment speeds for traffic increasing initially at 30% and then 15% annually. If no transition to digital occurs then the number of cells, and thus cost, escalates rapidly as each cell has very limited capacity. If an operator deploys CDMA more aggressively with a conversion of between 40% to 50% of traffic over to digital, then they can stop building additional cell sites altogether. At more gradual (moderate) rates of migration to digital, for example at 10% (25%) conversion in year 1 and 25% (50%) in year two, the growth in the required number of cells reduces, though the cells are now dual-mode and thus more expensive than pure analog cells. This later scenario is similar to what has occurred to CDMA deployments as a result of delays incurred procuring sufficient CDMA phones and the continued growth of analog. Similar conclusions had been arrived at by other cellular operators [e.g. 10]

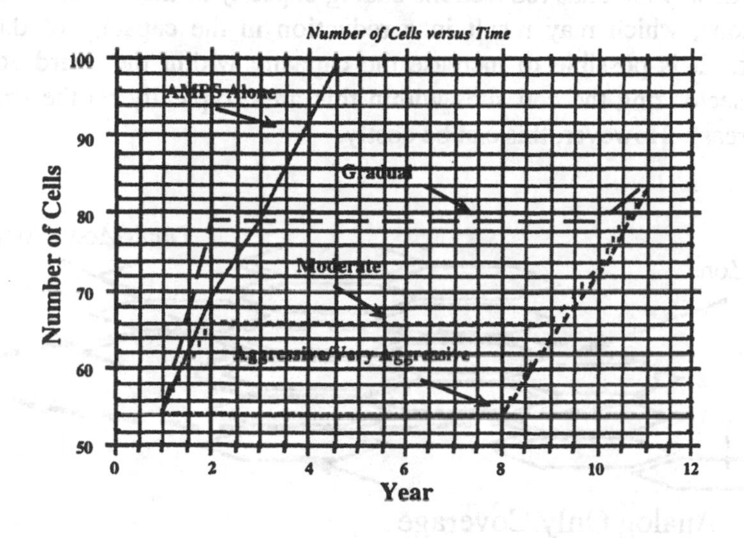

Figure 1. Number of cells required for different deployment Strategies as a function of time. See Reference [9] for background assumptions and detailed parameters used in the analysis.

Typically the CDMA system is not deployed over the entire analog network, but only along its most highly "trafficked" areas to form a dual-mode footprint within the much larger analog system, as illustrated in Figure 2. Since neither analog nor CDMA are immune to interference, this has resulted in a need for a geographical guard zone and guard bands between both systems. The guard zone is a geographical area, within which no analog cell site may use the frequency band assigned to CDMA. After a certain distance, measured in pathloss or interference terms, the spectrum may be re-used again by the analog system. The guard band refers to the AMPS frequency channels, which need to be cleared on either side of the CDMA carrier. Typically for one CDMA carrier the guard bands are 270 kHz (equivalent to 9 analog channels) on either side of the CDMA carrier. The constraints on the site's power amplifier, which is commonly used, for both the analog and digital transmissions, affect the power output for the CDMA pilot, traffic and other channels on the forward channel. All these constraints impact the practical capacity achievable of the combined dual-mode network [9]. This has resulted in operators experiencing growth in capacity within this footprint, but necessitating a reduction in the analog traffic carrying capacity in the area between (and within) the dual-mode footprint and the rest of the underlying analog network. Thus even though the capacity of the system within the footprint has grown, the effect of the CDMA deployment has reduced the analog capacity of the system within the guard zone, which may result in a reduction in the capacity of the total network. It is possible to increase the capacity within the guard zone by further sectorizing the cell sites within this area (typically on the fringe of urban areas). However, this can be costly.

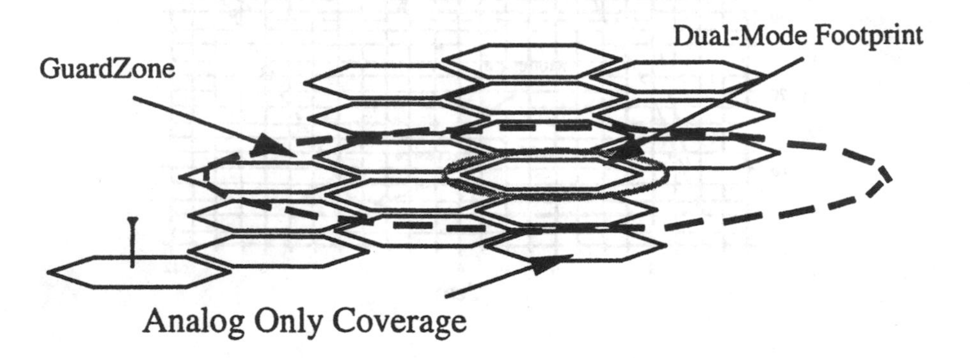

Figure 2. Illustration of a Dual-mode footprint, the guard zone and the underlying analog system

This poses a number of problems for the cellular operator. As the system grows one of the obvious choices is to go to additional carriers and effectively increase the CDMA capacity carrying capabilities of the network almost linearly. However to support the ever increasing Erlang traffic in currently deployed dual-mode networks, several options are available. This chapter discusses some preliminary results of the various options available and addresses the impact of increasing total traffic on the resulting analog system.

It is within this increased competitive market place where the legacy analog system is still important and digital migration cannot occur fast enough, that we deal with in this chapter. The question is how does one optimize the newly deployed CDMA network once deployed over the legacy analog network. This chapter discusses the issue of deploying CDMA over an incumbent analog system, how both systems can be optimized going forward and some future developments, which are envisioned to help continue to optimize the usage of the cellular spectrum, a scarce resource. Section 2 discusses the complex issues involved in deploying CDMA over an analog system. Section 3 concludes the chapter with a review of issues affecting CDMA deployment and some concluding remarks.

2. OPTIMIZATION OF THE TOTAL NETWORK

In this section we discuss how we can optimize the CDMA and analog network in an on-going basis. The options analyzed are:

- Increasing the overall capacity by appropriately increasing the number of dual-mode sites.

- The adoption of another carrier in the network and have some sites in the "hot spots" use this additional carrier. The resulting improvement in overall system capacity is weighed with the wider guard bands required at the AMPS sites lying in the guard zone. The third option studied is to add an additional CDMA carrier to the CDMA sites in the hot spots.

- The appropriate choice of CDMA-AMPS handdown parameter settings along the dual-mode/analog border. When a mobile user is moving out of the CDMA coverage area into the AMPS system, intelligent CDMA to AMPS Handdown (CAH) strategies must be employed so that the CAH regions are well defined and contiguous along the boundary area while keeping dropped calls to a minimum and maximizing CDMA capacity [11]. We analyze in depth these issues of how best to optimize the overall network so as to allow seamless transitioning between the analog and dual-mode networks by employing the above strategies and using the sector, or *cell size* parameter appropriately.

2.1 CAPACITY OF TOTAL NETWORK

CDMA Simulation

The analysis can be handled very easily by a simulation tool like GRANET[3], which simulates both the forward, and the reverse link as defined by IS-95 with the implementation outlined in references [12,13]. The propagation models are based on an enhanced Okumura-Hata model [14], where land-use and land-cover data provide clutter corrections, elevation data provides corrections for sloping terrain, etc. The scenarios considered in this paper are chosen to produce a pathloss of 32 dB per decade over a flat-earth model containing cells spaced 10 kilometers apart and at a height of 100 m. The underlying analog system is assumed to be transmitting at 50 Watts ERP.

A large number of parameters (both system wide and on a per sector basis) are available to the engineer for CDMA planning in real deployments. These have been incorporated into the simulation to help realize improvements in coverage projections. By these enhancements, we are better equipped to help plan the CDMA networks during deployment, which in turn implies more accurate predictions for the system performance.

We consider the case for a system operating at 1% FER on both the forward and the reverse links, at 9600 kbps[4], with a nominal traffic power of 1.75 Watts (+4.5 dB - 5.6 dB variation allowed due to power control), pilot at 6.3 W, sync power is set at 0.6 Watts, paging at 2.2 W. Each cell site is assumed to have a noise figure of 6 dB, cable loss of 2 dB and a 110 degree antenna with a nominal gain of 11.0 dBd for both the CDMA and analog cells. The dual-mode portables have a noise figure of 9 dB, maximum transmit power of 200 mW and an antenna gain of 0 dBd. We model the handoff process by considering a handoff threshold of -15 dB. This is equivalent to T_{drop}. A detailed description of the simulation tool and its features are described in reference [9,15].

As part of the dual-mode system we have addressed and modeled the necessary geographical area (guard zone) and spectrum (guard band) that must be cleared when deploying CDMA on an embedded AMPS network. This is necessary in defining the amount of spectrum that has to be cleared in the analog network as a function of the geographical distance from the

[3] GRANET is a registered trademark of GTE Laboratories Incorporated.

[4] The results are considered general enough to also apply to the higher rate vocoder at 14.4 kbps.

CDMA deployment area. This is heavily a function of the actual CDMA deployment and CDMA related parameters such as noise floor elevation and expected traffic carried by the CDMA system and the interference caused by the CDMA system. Algorithms were developed which quickly estimate the cell sites that require having part of their operating spectrum cleared.

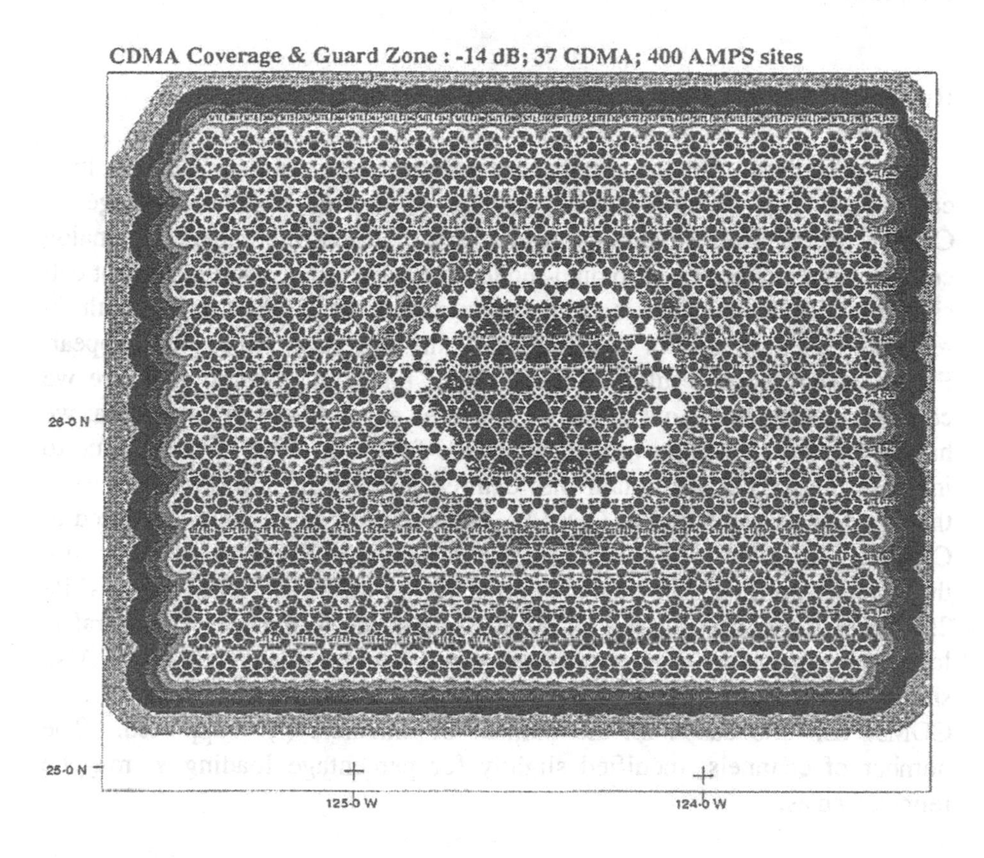

Figure 3. CDMA Coverage regions and guard zone as well as the underlying analog network. The different colors/shades represent different signal levels. The central "white" area along with the digital footprint represents the guard zone.

The simulation was run for 400 cell sites (tri-sectored) arranged in a regular hexagonal pattern, with 37 inner cell sites, itself forming a hex pattern, forming the dual-mode footprint as shown in Figure 3. Of these 37 cell sites that lie within the dual-mode footprint, only a portion were converted to CDMA as deemed necessary to carry the traffic loading while maintaining coverage within the footprint. The guard zone is represented by the " white region ", along with the digital footprint in Figure 3. The scenarios investigated were the effect on the traffic carrying capacity of the total system, including the size of the guard zone, as we convert more cell sites within the footprint to digital, and thus increase the digital carrying capacity. Secondly, the option of an additional carrier was investigated and also the effect of a non-uniform traffic distribution within the dual-mode footprint.

Results

As described above, initially the operator can deploy CDMA in a coverage-limited fashion, taking benefit of the coverage advantage of CDMA. As the traffic requirements increase, more of the underlying analog cell sites can be converted to dual-mode. If we do not convert sufficient cell sites to CDMA then the cell-breathing nature of CDMA will result in reduced coverage within the footprint and coverage holes will appear. Figure 3 shows the results for the scenario under consideration where we consider starting at 22 out of the 37 cell sites as CDMA and ensure that we have adequate CDMA coverage within the footprint. We then continue to increase the traffic and obtain the required number of additional cell sites that need to be converted to CDMA. The number of cell sites converted to CDMA to maintain the original footprint is shown in Figure 4, together with the traffic supported. The total supported digital traffic is 780 Erlangs, while 255 Erlangs is carried by the inner cells within the footprint. As the traffic load increases we need to convert more cell sites to dual-more sites. Also shown in Figure 3 is the theoretical calculation [4] for the capacity of a CDMA cell site based on the number of channels (*N*) supported. The number of channels, modified slightly for percentage loading χ may be represented as;

$$N = \frac{W}{R} \cdot \frac{1}{\frac{E_b}{N_o}} \cdot \frac{1}{d} . F . G . \chi \qquad (1)$$

where, $W = 1.23$ MHz is the signal bandwidth, R is the information rate, 9.6 kbps, Eb/No is set to 7 dB, d is the voice activity factor (0.5), F is the frequency reuse factor (that is the ratio of in-cell to out-of-cell interference, assumed equal to 0.6 here which is appropriate for the pathloss exponent employed in the simulation), G is the sectorization gain (assumed to be 2.55 for the directional antenna), χ can vary from 0 to 100%, with a realistic range of 50% - 75%. For the above values with χ of 50% we obtain an average of 30 Erlangs per cell, assuming channel pooling between the sectors and operating at 2% blocking.

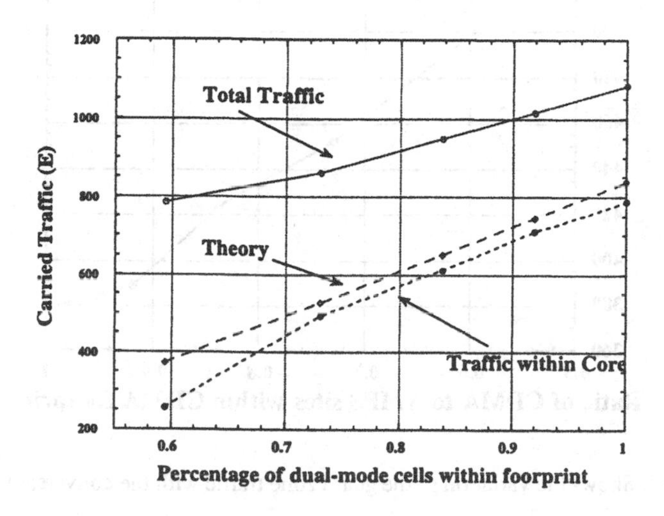

Figure 4. Variation of traffic with conversion rate (CDMA cells divided by the analog cells)

We consider the total traffic carried within the footprint core and also the total traffic carried by the system. The traffic carried by the inner-core of cells agrees well with theory, with the lower conversion ratios sustaining capacities below the theoretical level, due to being coverage limited rather than capacity limited. The total capacity of the footprint includes the traffic carried by the cell sites along the border of the footprint.

Initially when we deploy CDMA as cost effectively as possible over the underlying analog system, it is deployed as a coverage limited system and is more prone to interference (interference threshold assumed as -110 dBm) from the neighboring analog system. We therefore need to create a guard zone between the dual-mode system and the rest of the analog system, in which the analog system cannot use the 1.23 MHz that is required by CDMA, plus guard bands, resulting in approximately 1.8 MHz that has to be cleared. Figure 5 shows the variation of the guard band traffic with the conversion ratio (i.e. ratio of CDMA to AMPS sites within the digital footprint). At the lower conversion ratio (0.59), we have to clear approximately 500 sectors, reducing to approximately 380 for a conversion ratio of 1, when we deploy CDMA on every analog cell site within the footprint.

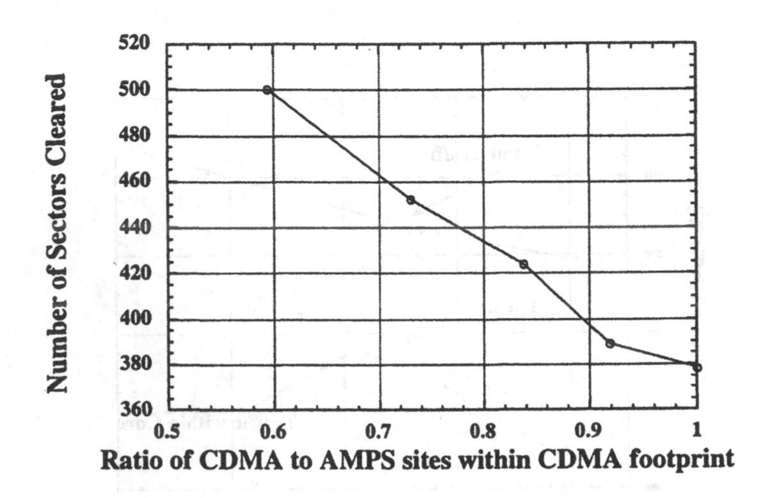

Figure 5. Shows the variation of the guard zone traffic with the conversion rate

However, as we increase the traffic the CDMA coverage starts to shrink. The coverage provided by the cell sites along the border starts to shrink and we are required to deploy more cell sites within the dual-mode footprint to maintain the same coverage. As the coverage along the borders shrinks slightly, the number of cells within the guard zone is also reduced. This reduces the impact of the CDMA system on the underlying analog system while increasing the traffic capacity of the analog system.

Table 1 illustrates the trend and shows the increase in the total traffic (AMPS + CDMA) with respect to the traffic in the dual-mode footprint and the size of the guard zone. As we increase the number of cell sites within the dual-mode footprint we can increase the traffic that we can carry. As the CDMA traffic increases we could opt to add an additional frequency to our CDMA system. If we go to an additional frequency we double effectively the CDMA capacity, though incurring an additional loss in the underlying analog system. From Table 1, we see that only when the conversion rate is around 0.84 or higher do we see that there is a benefit to going to another frequency.

Table 1. Increase in total Traffic (AMPS[5] + CDMA), with respect to the traffic in the dual-mode footprint and the guard zone.

Conversion Rate	CDMA Traffic	Sectors in Guard zone	Reduction in Analog Traffic (E)	Total Traffic (E)
0.59	785	500	827	(42)
0.73	861	452	748	114
0.84	949	424	701	248
0.92	1014	389	643	371
1.00	1084	378	625	459

Before the actual deployment uniform traffic amongst the cells is typically assumed. However, in a real deployment scenario, the traffic distribution will not be uniform, but will involve traffic hot spots. These hot spots will use the resources of the neighboring cell sites. We consider the baseline case of 27 CDMA cell sites within the dual-mode footprint (conversion factor = 0.73). As shown in Table1, the CDMA traffic carried was 861 Erlangs.

Table 2. Total traffic for deployment of an additional frequency, or deploying more cell sites

CDMA Frequency	# of dual-mode cells	Sectors in Guard zone	Reduction in Analog Traffic	CDMA Traffic (945 E)	Total Traffic (E)
1 2 (core only)	27 7	446 192	738 318	861 84 945	123 (234) (110)
1 2	27 27	446 446	738 738	472.5 472.5 945	(265) (265) (530)
1	29	452	748	945	197

To simulate the hotspot areas, the traffic in the center 7 dual-mode cells was increased by 4 Erlangs/sector. This led to coverage holes, due to the excess traffic around the center. The issue facing the service provider is

[5] AMPS is assumed to have nominally 18 channels per sector, at 2% blocking and CDMA to occupy 1.8 MHz of spectrum.

whether to deploy more cell sites in the neighborhood of these hot spots, or possibly go to an additional frequency. Table 2 shows the results of the 3 options. The first option of adding another CDMA carrier in the core led to a reduction in the total traffic of 110 Erlangs. The second option of using the additional carrier everywhere reduced the overall traffic that could be carried by 530 Erlangs. This is mainly as a result of the excessive loss in analog traffic capacity of 738 Erlangs. The third option in the scenario was to increase the number of cell sites within the dual-mode footprint. For this case it just takes an additional two sites to maintain the coverage objective and results in an additional 197 Erlangs being carried.

Of course as the traffic increases we can use the inherent potential of the additional carrier to allow for future growth. However, as seen in Table 2, the addition of another carrier is not always the best alternative. It is also clear that in defining the gain of CDMA over analog that the Erlang loss in traffic carrying capacity in the guard zone must also be taken into account.

2.2 Impact of CDMA-to-AMPS Handdown Procedures on Network Performance

Another important network optimization issue is the flawless network operation at the boundaries of the dual-mode (mainly urban areas) and analog only (typically more rural) areas. Mobile users in the border of a CDMA coverage area—or on the edge of a "CDMA coverage hole"—must be able to get their CDMA service gracefully handed down to the AMPS system. Careful network engineering and optimization is required at the borders so that the CDMA call is handed down to an explicitly specified AMPS server well before the CDMA signal gets too weak and the call is dropped. In a similar fashion, when there is a request for a call initiation in the CDMA-AMPS boundaries, it may be necessary for the call to be immediately handed over to the overlaid AMPS service.

To guarantee a flawless handdown procedure that is triggered and controlled by the network, a CDMA-AMPS handown (CAH) region should be identified at the boundary area between the CDMA and AMPS footprints. This region should be (1) contiguous, to guarantee a consistent handdown event along the CDMA-AMPS boundaries; and (2) not very wide, because this leads to a considerable loss in the traffic supported by the CDMA cells in the boundary of the dual-mode footprint. Typically, CAH strategies involve CDMA beacons, border cells, explicitly defined pilot neighbor lists for the sectors along the boundaries, and variable *cell sizes*[6] to facilitate a flawless CDMA to AMPS handover of service. We limit our studies here to the approaches based on border and beacon cells with variable *cell sizes*.

A border cell[7] is a regular CDMA server located in the CDMA-AMPS boundary area. It is equipped with the additional functionality of first triggering a CAH event and then transferring the CDMA call to the AMPS server that is usually collocated with the border server[8]. During a border-cell-assisted CAH, a mobile user is handed down to AMPS when both the following two conditions are met: (1) all CDMA pilots in the user's active set should be configured for this handdown type (i.e., they are border servers as well); and (2) the received CDMA pilot signal over the total interference (E_c/I_o) should be below a user-specified handdown threshold T_{brd} for all pilots in the user's active set.

A CDMA beacon is a pilot-only sector located in the dual-mode to AMPS geographical border area. It does not carry any traffic but is rather used to trigger a CAH event and maintain the necessary synchronization. During a CDMA beacon assisted handdown, a user is handed down to an AMPS server (usually collocated with the CDMA beacon) when at least one CDMA beacon with received pilot power (E_c/I_o) above a user-specified threshold T_{beac} has been detected.

The sector *cell size* is a tuning parameter typically available on every CDMA sector in the network. In deployed networks, due to the coverage versus capacity tradeoffs, the serving areas of CDMA cells can vary dynamically in size with the changing traffic conditions. Typical base station equipment can be programmed with a distance/time related parameter referred to as *cell size* to limit undesirable soft-handoffs and call initiations from "weak" portables. A proper setting of the *cell size* helps eliminate "rogue" pilots from entering the active set of a portable, especially from base stations located far away with very tall antennas, or in areas of potential "pilot-pollution" where the portable may have to change the status of its active set much too often. *Cell size* is also important in facilitating intelligent CAH strategies involving CDMA beacons. A CDMA-to-AMPS handdown event that is assisted via the *cell size* parameter, denoted as C in our analysis,

[6] This is a sector parameter which limits the distance within which the sector's pilot will be considered as a possible server or a handoff candidate.

[7] By "border cell," we mean a sector deployed in a regular CDMA base station which is on the edge of the dual-mode network and its primary function is to seamlessly hand-down to analog gracefully.

[8] This is usually the case as it reduces confusion as to which analog carrier the signal should be handed down too.

at a CDMA beacon, necessitates that a user is handed down to an AMPS server when (a) at least one CDMA beacon with received pilot power (E_c/I_o) above a user specified threshold T_{beac} has been detected and (b) the user's distance from the corresponding beacon is within the range C as defined by the value of the *cell size* parameter set on that beacon.

We have conducted numerous simulations to study the impact of the aforementioned CAH strategies on the performance of a dual-mode CDMA network. The performance criteria used were the average capacity loss and average handoff overhead per boundary cell. The average capacity loss per boundary cell represents the loss in CDMA primary traffic carried by a boundary cell with border or beacon sectors relative to the CDMA primary traffic carried by the same boundary cell without any borders or beacons. The average handoff overhead per boundary cell represents the ratio of total CDMA traffic carried by a boundary cell over the CDMA primary traffic carried by the same cell. It defines the additional number of channel elements required to support handoff (virtual) traffic and provides an indication of network efficiency. The effect of the tuning parameters T_{brd}, T_{beac} and C on the system performance is described next. Additional details may be found in [9,14].

The simulation is as described previously with similar parameters except as noted below. The CDMA network considered is loaded at approximately 40% of its pole capacity. The voice coding with overhead is assumed at 14.4 kbps (rate set II). The nominal traffic channel power is set at 2.7 W with a pilot channel power of 5.0 Watts. The remaining paging and sync control channels have a total power of 2.2 Watts. Every cell site is assumed to have a noise figure of 6 dB, cable loss of 2 dB and a 110 degree antenna with a nominal gain of 11.0 dBd for both the CDMA and analog cells. One CDMA channel is used from the cellular band centred at 881.52 MHz. CDMA soft handoff parameters (T_{add}, T_{drop}) are set at negative 15 dB, and a T_{tdrop} of 2.5 seconds is assumed.

CDMA Border-Cell-Assisted Handdown

We investigate the performance tradeoffs in network coverage, capacity and handoff with respect to the number of border sectors deployed per boundary site and the handdown threshold T_{brd}. One, two, or three border sectors are deployed per boundary cell site and T_{brd} varies from –2 to –6 dB. Figure 6(a) shows the variation of the average capacity loss versus the handdown threshold T_{brd}. As the handdown threshold increases, the average loss in capacity also increases. This is because the probability of having CDMA border pilots in the active set decreases with an increase in T_{brd}, resulting in more CAH events and hence a loss in CDMA traffic at the boundary cells.

In addition, the loss in capacity increases as the number of deployed border sectors per boundary cell increases. When one border sector is deployed, no more than 20% of the offered CDMA traffic at the boundary cell is handed down to AMPS. However, a contiguous CAH area cannot be formed around the dual-mode footprint. The CAH area becomes contiguous when two border sectors per boundary cell are deployed and the capacity loss in this case is up to 50%. Finally, most of the CDMA traffic offered at the boundary cells is handed down to analog when three border sectors per boundary cell are deployed. However, in this case, the CDMA traffic loss may be as high as 90%, causing the CDMA coverage to shrink by almost one tier of cells.

Figure 6(b) shows the impact of handdown threshold T_{brd} on the average handoff overhead at the boundary cells. As the handdown threshold increases, an increase in the handoff overhead is also observed; this increase is more apparent in the case of three border sectors per cell. For the other two cases, the impact of T_{brd} on the handoff overhead is not significant and is similar to that observed in the case of no border sectors at the boundary cell.

CDMA Beacon-Assisted Handdown

Alternatively, CDMA beacons were deployed either at the boundary CDMA site or at the AMPS sites within the first tier of analog network surrounding the CDMA footprint. The impact on the loss in capacity and handoff overhead was studied by varying the handdown threshold T_{beac} from –4 dB to –16 dB. Figure 7 (a) shows the variation of the average capacity loss with the handdown threshold T_{beac}. As observed, a decrease in handdown threshold leads to an increase in the average capacity loss. This is because the probability of having CDMA beacon pilots in the active set increases with a decrease in T_{beac} resulting in more CAH events and hence a loss in CDMA traffic at the boundary cells.

At thresholds below –12 dB, there is no impact on the amount of traffic handed down to analog because the interference (I_o) at the boundary is lower than that at the core of the CDMA footprint. The CAH strategy of placing two beacons per cell leads to the highest loss in capacity on the order of 80%, causing the CDMA coverage to shrink by almost one tier. On the other hand, placing beacons at the first tier of AMPS cells leads to the minimum loss in capacity. In particular, thresholds of the order of –8 dB lead to a minimum loss in capacity while maintaining a contiguous CAH area. For thresholds below –8 dB, a contiguous CAH area is maintained for all three cases.

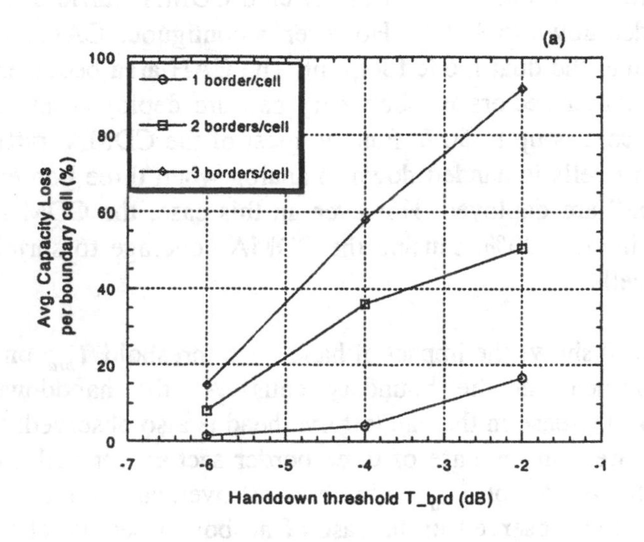

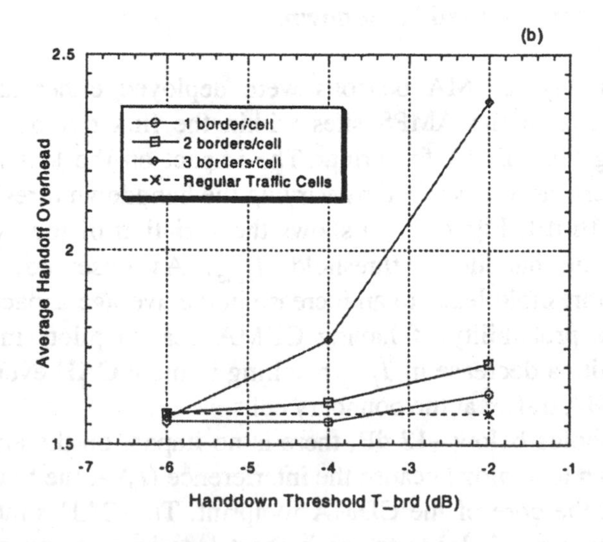

Figure 6. Impact of Handdown threshold T_{brd} on performance of a dual-mode network at its boundaries with an analog system. (a) The average capacity loss per boundary cell site and (b) the average handoff overhead per boundary cell site

Figure 7 (b) shows the impact of handdown threshold T_{beac} on the average handoff overhead at the boundary cells. Similarly with the average capacity loss, a decrease in the handdown threshold leads to an increase in the handoff overhead. Lowest handoff overheads are observed in the case when one beacon sector is deployed at the AMPS sites outside the CDMA footprint. In this case, the overhead is even lower than that observed when these are no beacon sectors at the boundary cell. The impact of T_{beac} at values above –12 dB is more apparent for CAH strategies involving one or two beacons at the CDMA boundary cells.

Cell Size - Assisted Handdown

The last CAH strategy investigated involves the deployment of beacon sectors with limited cell sizes at the network boundaries. Two sub-cases are considered depending on the number of beacons (one or two respectively) deployed on every CDMA site located at the boundary tier of the CDMA network. To reduce the large search space, the handdown threshold T_{beac} was fixed at –15 dB. Figure 8 shows the impact of varying cell size to the average CDMA traffic loss per boundary cell and the associated handoff overhead. As the *cell size C* increases, the CDMA traffic carried by the outer sites decreases, or equivalently, the capacity loss increases. This lost CDMA traffic is handed down to AMPS. This is expected, since a large cell size defines a larger CAH region over which users are handed down to AMPS when they detect a strong beacon pilot. In the case of one beacon deployed per boundary site, the CDMA capacity loss increases from 15% to 57% as C is varied from 0.5 to 1.5. A similar trend is observed when two beacons are deployed per boundary site. However, the CDMA traffic carried in the latter case is significantly less when compared to that of the former. This is expected since only one sector in these sites is equipped to carry CDMA primary traffic. Another observation is that when C is lower than 1, the CAH region does not remain contiguous anymore. This would be undesirable since it does not facilitate graceful CDMA to AMPS handdown events over the entire CDMA boundary. For higher values of C, CAH contiguity is maintained although the CDMA traffic carried is significantly reduced as illustrated before.

The cell size also affects the handoff overhead at the boundary sites. The two-beacon per site deployment scenario appears to be very sensitive to variations in C; even a small increase in C makes the handoff overhead high. On the other hand, variations in C appear to have a minimal impact on the handoff overhead when one beacon per site is deployed. In summary, a cell size on the order of the cell radius provides a contiguous CAH region while maximizing the CDMA traffic carried and keeping the handoff overhead low.

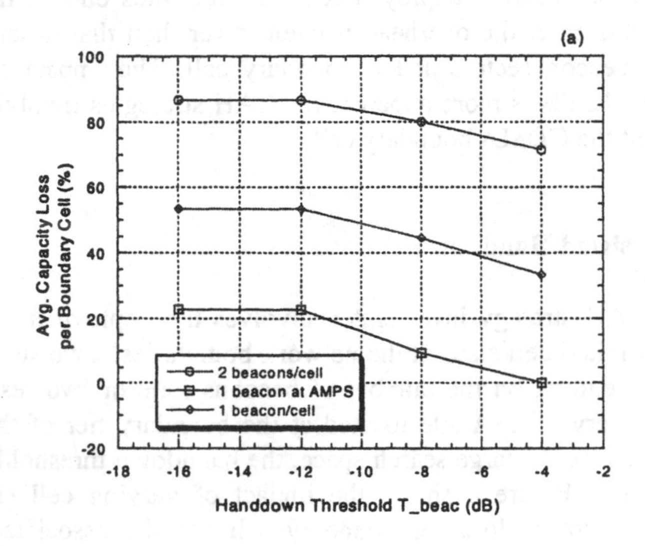

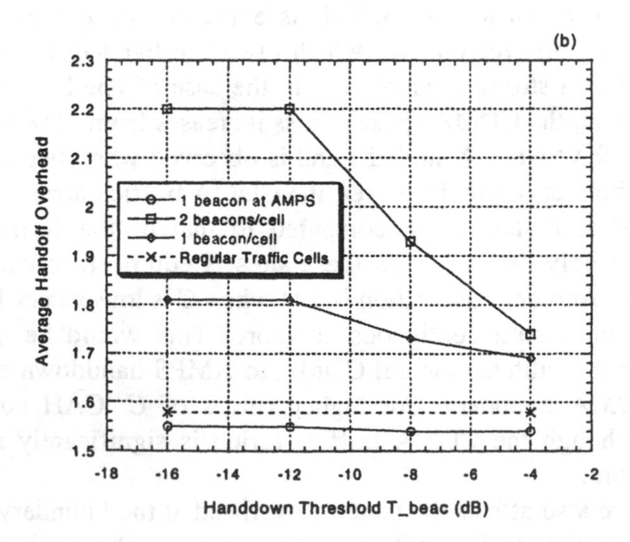

Figure 7. Impact of handdown threshold T_{beac} on the performance of a dual-mode network at its boundaries with the analog system. (a) shows the average capacity loss per boundary cell site and (b) the average handoff overhead per boundary cell site

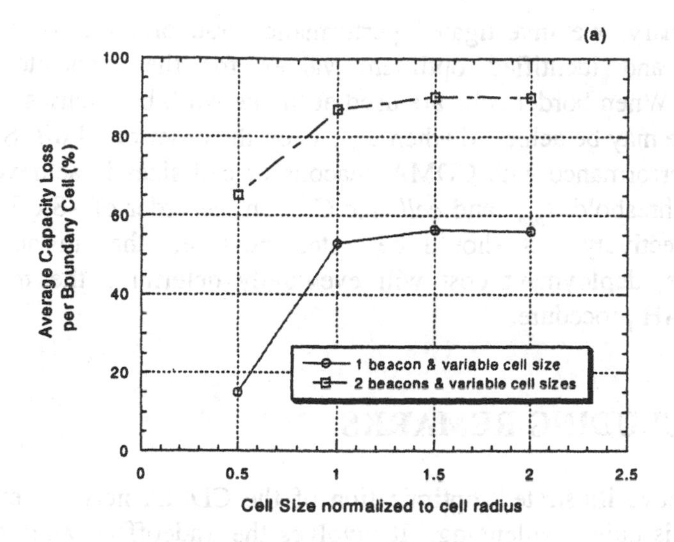

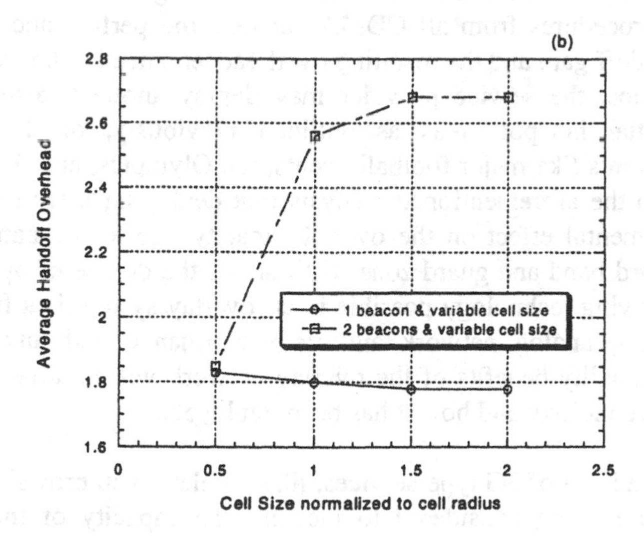

Figure 8. **Impact of the cell size C on (a) the average CDMA capacity loss and (b) handoff overhead with one or two beacon sectors deployed at the boundary cells**

In summary, we investigated performance tradeoffs for various CAH procedures and identified optimum values for the associated tuning parameters. When border cells are used at the network boundaries, optimum performance may be achieved when T_{brd} is on the order of –4 dB. Similarly, optimum performance with CDMA beacons or cell sizes is achieved when handdown threshold T_{beac} and *cell size C* is on the order of –8 dB and cell radius respectively. It should be noted however, that in addition to performance, deployment cost will eventually determine the most cost-effective CAH procedure.

3. CONCLUDING REMARKS

As we have illustrated, optimization of the CDMA network even with one carrier is quite challenging. It involves the tradeoff between capacity, coverage and voice quality. With the constant increase in traffic demand, service providers are deploying additional CDMA carriers. When another carrier is added in the network, issues concerning inter-carrier handoff, handdown procedures from all CDMA carriers, the performance tradeoffs between handoff gain and the resulting overhead, etc. need to be explored. In some scenarios, the service provider may deploy another carrier only in some potential hotspot areas as outlined previously, or during some temporary events like major football events, the Olympics, etc. However, it is clear from the aforementioned analysis that deploying more carriers can have a detrimental effect on the overall capacity due to the nature of the resultant guard band and guard zone. Ultimately, the degree of optimization by the underlying technology possible in an overlay system is a function of the underlying analog network and its performance and any capacity, coverage or quality benefits of the overlay network will be affected by the underlying technology and how it has been deployed.

With the advent of 3G type services, like wireless web browsing, several techniques are being considered to increase the capacity of the network while maintaining a consistent QoS. One of the techniques is the inclusion of a smart antenna system. Such systems typically include multi-beam or adaptive beam forming and have been shown to provide capacity improvements in digital cellular networks [16]. Other techniques include Tower-Mounted Amplifiers (TMA) and/or super-conducting filters [17] which can provide several dB's improvement in reverse link coverage, which is sufficient in many cases to correct coverage holes. TMA's have not been much used in the past due to maintenance and lightning concerns, but their application needs to be re-evaluated as a valuable alternative to adding digital-only cell sites. It is thus imperative that the choice of migration technology is not based on the ultimate performance of said technology but of the total deployed network and off the synergies that can be fostered between the technologies.

REFERENCES

[1] V. H. McDonald," The Cellular Concept", BSTJ, no. 58, pp. 15-40, 1979.

[2] W. C.Y. Lee," Mobile Cellular Telecommunications", McGraw-Hill New York, 1989.

[3] TIA/EIA/IS-95, "Mobile Station-Base Station Compatibility Standard for Dual-Mode Wideband Spread Spectrum Cellular System," *Telecommunications Industry Association (TIA)*, Washington, DC, 1993.

[4] A. J. Viterbi and A.M. Viterbi, "Erlang Capacity of a power controlled CDMA system'" *in IEEE Selected Areas in Communications*, pp. 892-900, Aug. 1993.

[5] S. Titch, "Blind Faith", *Telephony*, Sept. 8, 1997.

[6] M. Riexenman," Communication", IEEE Spectrum, pp. 29-36, Jan. 1998.

[7] E. Jugl, H. Boche," Limits of Sectorization gain caused by Mobility and soft handoff", *Elect. Letts*. Pp. 119-120, vol. 35, #2, Jan. 21, 1999.

[8] Ameritech," ClearPath™ Media Kit", http://www.ameritech.com.

[9] R. Ganesh and V. O'Byrne, "Improving System Capacity of a Dual-Mode CDMA Network," *Proc. of IEEE ICPWC*, pp. 424–428, 1997.

[10] H. Chan and C. Vinodrai," The Transition to Digital Cellular", 40th Vehicular Technology Conference, pp. 191-194, Orlando FL, May 1990.

[11] H. Stellakis and R. Ganesh, "CDMA to AMPS Handdown Strategies In a Dual-mode Cellular Network", in *Proceedings of Int'l Conference on Communications (ICT'98)*, Greece, June 1998.

[12] H. Stellakis and A. Giordano, " CDMA Radio Planning and Network Simulation", in *Proc. IEEE Int. Symposium. On Personal, Indoor and Mobile Communications*, Taiwan, 1996.

[13] M. Wallace and R. Walton, "CDMA Radio Network Planning," *Proc. of IEEE ICUPC*, pp. 62–67, 1994.

[14] M. Hata, "Empirical Formula for Propagation Loss in Land Mobile Radio Services", *IEEE Transactions on Vehicular Technology*, vol. VT-29, Aug. 1980.

[15] R. Ganesh, H. Stellakis, "Impact of cell size on capacity and handoff in deployed CDMA networks", Elect. *Letts*, pp. 2205-2207, Vol. 34, No. 23, Nov. 1998.

[16] Y. Li, M. J. Feuerstein, D. O. Reudink, "Performance Evaluation of a Cellular Base Station Multi-beam Antenna", *IEEE Transactions on Vehicular Technology*, Vol. 46, Feb. 1997.

[17] Marc Rolfes, "Speeding the cell-site acquisition process", Mobile Radio Technology, November 1996.

Chapter 5

MICROCELL ENGINEERING IN CDMA NETWORKS

DR. JIN YANG

Vodafone AirTouch Plc.

Abstract: CDMA microcell engineering is systematically studied and presented. The embedded microcell shares the same frequency and has full connectivity with the overlaying macrocell. The capacity of microcell and macrocell is derived and simulated at various traffic distributions. The microcell capacity is 1.03 to 1.12 times the capacity of a regular cell. The capacity of the combined microcell and macrocell is 2.00 to 2.11 times that of a regular cell. The microcell and macrocell performance is also analyzed in terms of RF reliability, soft hand-off factors, interference and power levels. The macrocell RF reliability will degrade more seriously than that of the microcell. The radio hand-off factors of the microcell are about 11% higher than that of the macrocell. The average required forward traffic channel power of the microcell is about 10% less than that of the macrocell. Microcell engineering guidelines in a commercial CDMA system are also provided. The results show that embedding the microcell in an existing CDMA network could be a very efficient way to improve hot-spot capacity and dead-spot coverage.

1. INTRODUCTION

Microcells play an important role in expanding an existing wireless network. Code Division Multiple Access (CDMA) communications systems have been deployed and commercialized all around the world [1]. The traffic demand for CDMA wireless systems has grown rapidly in recent years. The emerging next generation wireless systems are hierarchical systems to serve customers' various needs.

Wireless networks consist of various layers of macrocells, microcells and picocells. The macrocell service area radius is usually larger than 1 kilometer. The microcell service area radius is around several hundred meters. The picocell service area radius is less than 100 meters. The underlay-overlay hierarchy has been used in frequency division multiple access wireless systems to achieve high capacity. The underlay system will transmit at a different frequency from the overlay system. There is little power co-ordination among microcells and macrocells. However, complex frequency planning and re-tuning are needed for the entire network. Hard hand-off boundaries are also difficult to define.

In a CDMA system, users of one CDMA carrier are sharing the same frequency bandwidth. This poses the challenge to implement a macrocell/microcell hierarchical structure [2] [3] among cell sites that use the same CDMA carriers. The hierarchical cell sites must effectively control their powers to eliminate mutual interference. Soft hand-off among microcells and macrocells reduces the required transmit power around cell boundaries. It can also easily adapt to irregular cell boundaries. Frequency planning is not needed.

There are several types of macrocell/microcell structures:

I. Macrocell/microcell with coverage defined by natural environments: for example, a mountain top macrocell covers a large area with around 10 mile radius, while several microcells inside the macrocell coverage area cover major traffic spots, such as along highways. This type of microcell has a regular cell site configuration and set-up.

II. A minimalist version of a microcell with remote low power RF transceivers: for example, the RF radiated signals are converted to optical signals, and then relayed through optical fiber to an attended center cell site. This type of microcell consists of two special components: the optical conversion unit and the low power RF radiator. It is typically used for tunnels, subways, or in-building coverage, where remotely transmitted low power is necessary to achieve the coverage. The microcell main unit consists of baseband processing and network interface functions. The remote RF units contain transmit power amplifiers and receiver pre-amplifiers, which interface with the antennas. The microcell main unit can serve one or more RF units through optical fiber or coaxial cable. Some of the microcells may feed a leaky cable or distributed antenna system.

III. Simulcast of macrocell and several microcells: A high powered sector simulcasts with microcell transceivers, or several microcell transceivers simulcast together. This type of structure requires high isolation among coverage areas of simulcasting cell sites.

The Type II microcell discussed above is a special case of Type I. Remotely located low power transceiver units can significantly reduce network cost. However, the time delay on optical fiber, or other transmission media, needs to be taken into account. Search window sizes and other parameters may need to be adjusted.

Most of the studies on heterogeneous cells are either based on partially isolated microcell and macrocell tele-traffic [3], or no soft hand-off between the microcells and macrocells [4]. Microcell capacity analyses were limited to the reverse link [2]. The performance impacts of additional CDMA cell sites on the existing network are still not well understood.

In this chapter, a practical microcell structure based on IS-95 CDMA systems is systematically studied on both the forward link and the reverse link. The embedded microcell shares the same frequency and has full connectivity with the overlaying macrocell. Soft hand-off is applied among microcells and macrocells. The required transmit powers for both the macrocell and the microcell are controlled to the minimum levels, using the AirTouch radio planning tool. The impacts of various traffic densities and cell site distances between microcells and macrocells are presented. The microcell and macrocell performance is also analyzed in terms of RF reliability, soft hand-off factors, interference and power levels.

Section 2 presents the microcell and macrocell propagation models used in this chapter. Section 3 evaluates system capacity and performance. Section 4 presents microcell simulation results. Section 5 provides microcell engineering guidelines. Section 6 presents the conclusions.

2. MICROCELL AND MACROCELL PROPAGATION MODELS

Microcell sites are usually used for hot traffic spots, or small coverage holes. Fig. 1 shows a typical macrocell/microcell geometry.

The Type I microcell has a complete base station unit located in the microcell site. The Type II microcell will have only a RF transceiver unit located in the microcell site connected to the base station unit located in the overlay macrocell. Compared to a macrocell, a microcell is usually more flexible and designed to be deployed with a small size and light weight. It could be mounted on a wall, pole or floor. A weatherproof cabinet is usually used for an unattended microcell.

In a mobile communications environment, the radio propagation pathloss (denoted as *T(y)*) can be modeled as

$$T(y) = g(Hs, hm, \xi) \cdot y^{-\alpha} \quad (1)$$

where y denotes the distance between the base station and the mobile station.

α denotes the pathloss exponent factor. The corresponding pathloss slope is 10α dB/decade.

g(.) represents the combined impacts of antenna heights (base station antenna height *Hs* and mobile station antenna height *hm*), terrain and environment impacts (ξ), etc.

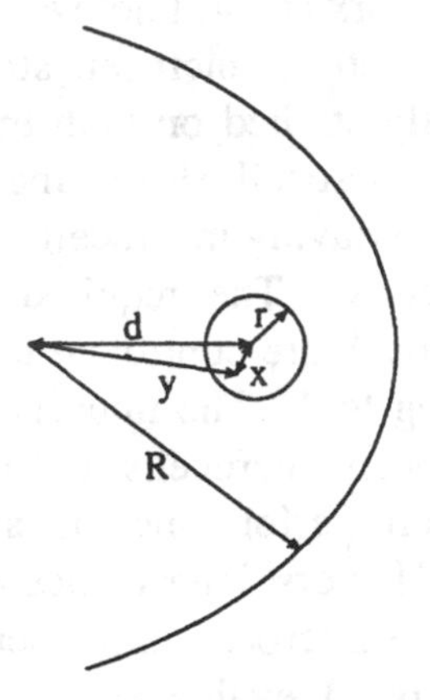

Fig. 1. Typical Macrocell/Microcell Geometry

A typical suburban propagation model is used for the macrocell. The microcell propagation model was derived from empirical data collected in a suburban neighborhood in San Francisco [5]. Fig. 2 shows the received signal strengths as measured for the macrocell and the microcell. Note that the microcell is accommodated inside the macrocell. The outward boundary of the microcell stretches farther out because the difference in received signal strengths is smaller. The soft hand-off zone is also stretched farther out, as shown in Fig. 3. A directional microcell with the antenna towards the overlay cell site could achieve more centralized microcell coverage.

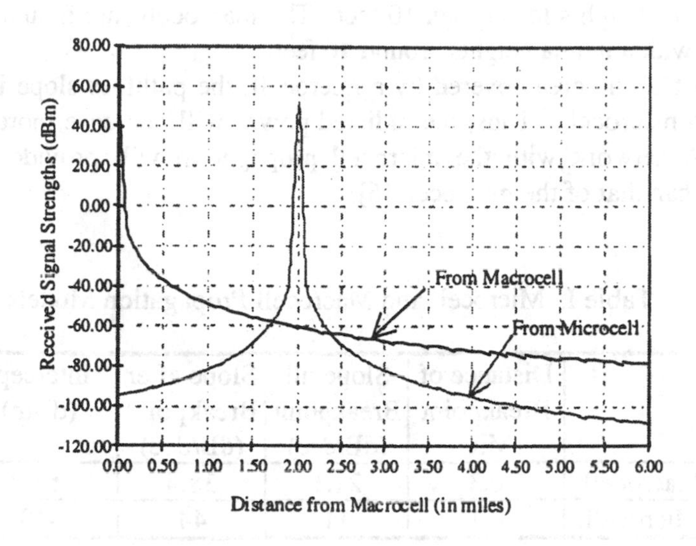

Fig. 2. Macrocell/Microcell Received Signal Strengths

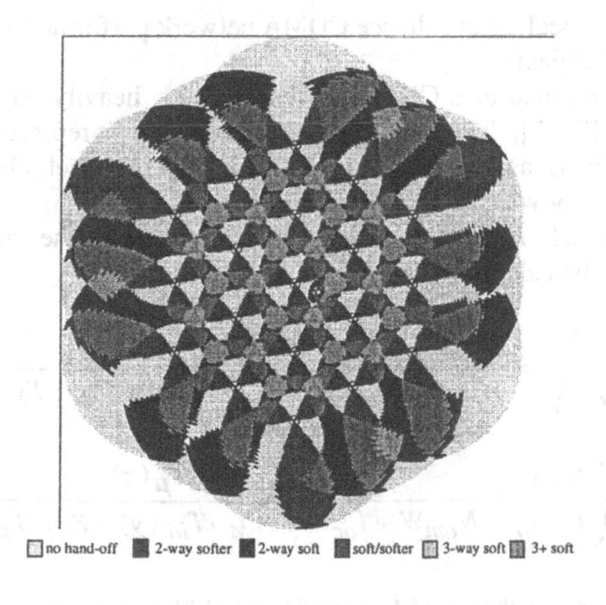

Fig. 3. Soft Hand-off Plot of the Simulated System

Table I below shows the slopes and intercepts used for the microcell and macrocell propagation models. The microcell is mounted under rooftops with antenna heights lower than 10 feet. The macrocells are mounted above rooftops, with antenna heights around 40 feet.

In most of the area covered by a microcell, the pathloss slope is steeper than for a macrocell. Thus, the radiated power will decrease more rapidly. A CDMA network with the microcell propagation will provide a higher capacity than that of the macrocell [6].

Table I. Microcell and Macrocell Propagation Models

	Distance of Breakpoint (Mile)	Slope till Breakpoint (dB/dec)	Slope after Breakpoint (dB/dec)	Intercept (dBm)
Macrocell	0.3	21.1	38.4	-61.7
Microcell	0.1	11	44	-40

3. CAPACITY AND PERFORMANCE EVALUATION

Use of microcells can enhance CDMA network performance and increase coverage and capacity.

The performance of a CDMA system depends heavily on pilot channel performance [7] [8]. The pilot channel performance is represented by pilot E_c/I_o, defined as the ratio of received pilot channel chip energy to received signal power spectral density.

The received E_c/I_o at a mobile station from the macrocell and microcell can be calculated as

$$\left(\frac{E_c}{I_o}\right)_M = \frac{\zeta_{pM} \cdot P_{cM} \cdot T_M(y)}{N_{om}W + I_{oc} \cdot W + P_{c\mu} \cdot T_\mu(x) + P_{cM} \cdot T_M(y)} \quad (2)$$

$$\left(\frac{E_c}{I_o}\right)_\mu = \frac{\zeta_{p\mu} \cdot P_{c\mu} \cdot T_\mu(x)}{N_{om}W + I_{oc}W + P_{cM} \cdot T_M(y) + P_{c\mu} \cdot T_\mu(x)} \quad (3)$$

where W denotes the CDMA system bandwidth.

$N_{om}W$ represents the non-CDMA RF background noise power.
$I_{oc}W$ denotes the interference from other CDMA cells in the system, excluding the macrocell and microcell of interest.
P_{cM} represents the total radiated power from the macrocell.
$P_{c\mu}$ denotes the total radiated power from the microcell.
ζ_{pM} and $\zeta_{p\mu}$ represent the pilot channel power percentage from the macrocell and the microcell, respectively.
$T_M(y)$ and $T_\mu(x)$ are the transmission pathloss from the macrocell and microcell sites to a desired spot, respectively. The desired spot is located at distance y from the macrocell site and distance x from the microcell site, as shown in Fig. 1.

The forward link performance is also determined by the traffic channel performance. The forward link performance is defined as the ratio of forward link received bit energy to noise power spectral density. Based on (2) and (3) above, the E_b/N_o from the macrocell and microcell can be derived as

$$\left(\frac{E_b}{N_o}\right)_M = \frac{\zeta_{tchM} \cdot P_{cM} \cdot T_M(y) \cdot PG}{N_{om}W + I_{oc}W + P_{c\mu}T_\mu(x) + P_{cM}T_M(y)(1-\zeta_{tchM})(1-\rho_M)} \tag{4}$$

$$\left(\frac{E_b}{N_o}\right)_\mu = \frac{\zeta_{tch\mu} \cdot P_{c\mu} \cdot T_\mu(x) \cdot PG}{N_{om}W + I_{oc}W + P_{cM}T_M(y) + P_{c\mu}T_\mu(x)(1-\zeta_{tch\mu})(1-\rho_\mu)} \tag{5}$$

where ζ_{tchM} and $\zeta_{tch\mu}$ represent the traffic channel power allocation percentage for the macrocell and microcell of interest, respectively. PG denotes the processing gain of the spread spectrum system. ρ_M and ρ_μ represent the orthogonal factor of the macrocell and microcell, respectively. The orthogonal factor is defined as a degradation of the signal due to the inability of rejecting other user interference in the forward links of the same cell. These include imperfections in the transmitter and receiver chains as well as multipath interference. An orthogonal factor of 1 means purely orthogonal and of 0 means totally lost orthogonality.

Defining p as the ratio of the received signal strength from the microcell to that from the macrocell at the mobile station, then

$$p = \frac{P_{c\mu} \cdot T_\mu(x)}{P_{cM} \cdot T_M(y)} \tag{6}$$

Combining (2) and (3), we have

$$\left(\frac{E_c}{I_o}\right)_\mu \Big/ \left(\frac{E_c}{I_o}\right)_M = p\frac{\zeta_{p\mu}}{\zeta_{pM}} \tag{7}$$

And

$$\left(\frac{E_c}{I_o}\right)_\mu = \frac{\zeta_{p\mu} \cdot p}{(N_{om}W + I_{oc} \cdot W)/(P_{cM} \cdot T_M(y)) + 1 + p} \tag{8}$$

That is,

$$\left(\frac{E_c}{I_o}\right)_\mu = \frac{\zeta_{p\mu}}{\zeta_{tch\mu}\left[PG/(E_b/N_o)_\mu + 1 - \rho_\mu\right] + \rho_\mu} \tag{9}$$

Thus, to achieve a target E_b/N_o under the specified E_c/I_o environment, the required traffic channel power allocation percentage is calculated as

$$\zeta_{tch\mu} = \frac{\zeta_{p\mu}/(E_c/I_o)_\mu - \rho_\mu}{PG/(E_b/N_o)_\mu + 1 - \rho_\mu} \tag{10}$$

The above formula can also be applied to a regular cell site configuration. For example, when the pilot channel power percentage is 15%, the received E_c/I_o is –9 dB, the processing gain is 85, and the orthogonal factor is 0.8. A traffic channel power allocation of 4.5% is required to achieve a target E_b/N_o of 10 dB.

Fig. 4 shows the relationship between the required forward link traffic channel power allocation and E_c/I_o operation points. When E_c/I_o increases, the required traffic channel power decreases. Within the microcell, the microcell E_c/I_o is larger than that of the macrocell, the microcell traffic channel power is smaller than that of the macrocell.

The cell boundary between the microcell and the macrocell is defined as the intersection of the transmission curves in Fig. 2. Around the middle of the cell boundary, the forward link pilot E_c/I_o from both cell sites should be equal. Therefore, we have

$$\frac{\zeta_{p\mu} \cdot P_{c\mu} \cdot T_{\mu}(x)}{\zeta_{pM} \cdot P_{cM} \cdot T_{M}(y)} = 1 \tag{11}$$

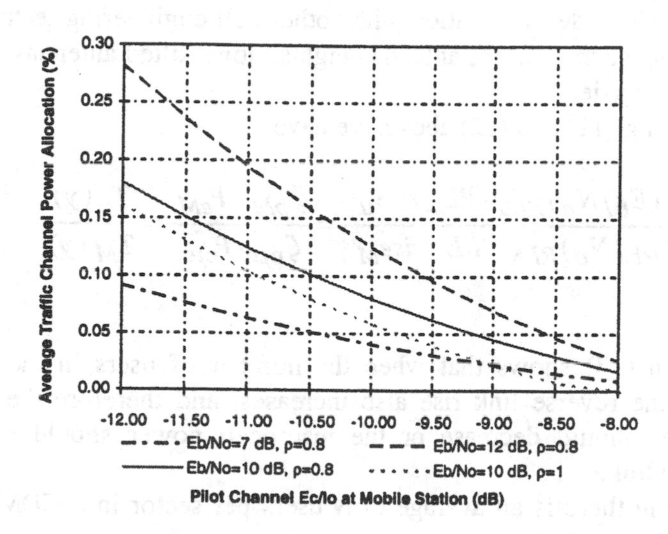

Fig. 4. Traffic Channel Power Allocation versus Pilot E_c/I_o

Since the transmission loss from the microcell is less than that from the macrocell, the pilot transmit power from the macrocell should be set to a higher level than that of the microcell.

To balance the reverse link cell boundary with that of the forward link, the required transmit power from the mobile station should also be the same, i.e.,

$$\frac{(E_b/N_o)_{RL\mu} \cdot RL\_rise_{\mu}}{T_{\mu}(x)} = \frac{(E_b/N_o)_{RLM} \cdot RL\_rise_{M}}{T_{M}(y)} = \frac{P_m \cdot PG}{N_o W} \tag{12}$$

where RL_rise_M and RL_rise_{μ} denote the ratio of total interference to the noise power at the macrocell and microcell of interest, respectively.

P_m denotes the mobile station transmit power.

$(E_b/N_o)_{RLM}$ and $(E_b/N_o)_{RL\mu}$ represent the required reverse link E_b/N_o at the macrocell and microcell sites, respectively.

The transmission loss from the microcell is less than that from the macrocell. For the same required E_b/N_o at both cell sites, (12) shows that the microcell reverse link rise is less than that of the macrocell. Therefore, when the microcell loading is the same as the macrocell loading, an attenuation pad may be added on the microcell receiver to balance the link. This can achieve desensitization when other cell engineering techniques are not available, such as lower antenna heights, down-tilted antennas, or natural propagation barriers.

Combining (11) and (12) above, we have

$$\frac{(E_b/N_o)_{RL\mu}\cdot RL\_rise_{\mu}}{(E_b/N_o)_{RLM}\cdot RL\_rise_{M}}=\frac{\zeta_{pM}\cdot P_{cM}}{\zeta_{p\mu}\cdot P_{c\mu}}=\frac{T_{\mu}(y)}{T_M(y)} \tag{13}$$

Equation (13) shows that when the number of users in the microcell increases, the reverse link rise also increases, and therefore the microcell pilot power should decrease or the macrocell power should increase to balance the link.

Assuming there is an average of N users per sector in a CDMA system with S sectors, we have

$$\left(\frac{E_c}{I_o}\right)_{\mu}=\frac{\zeta_{p\mu}\cdot p}{\dfrac{N_{om}W+\sum_{i=1}^{S}(\zeta_{pps_i}+N\zeta_{tch_i})P_{c_i}T_i(z)\cdot W}{P_{cM}T_M(y)}+1+p} \tag{14}$$

where ζ_{pps_i} and ζ_{tch_i} represent the combined pilot, paging and synchronization channel power percentage and the average traffic channel power percentage of the i'th sector, respectively.

P_{c_i} denotes the total transmit power from the i'th sector.

$T_i(z)$ denotes the transmission pathloss from the i'th sector.

When the number of users in the system increases, the E_c/I_o decreases, and the required traffic channel power will increase. However, when the E_c/I_o decreases, the soft hand-off area shrinks accordingly.

In a commercialized CDMA network, system capacity data must be obtained from a network performance monitor. A practical way to calculate capacity is based on power and interference measurement data from the forward and reverse links.

On the forward link, the transmit power percentage of various channels can be obtained. The CDMA forward link capacity (N_{FL}) can be calculated as

$$N_{FL} = \frac{1-\zeta_{pilot}-\zeta_{paging}-\zeta_{sync}}{\nu \cdot H_{rf} \cdot \zeta_{tch}} \tag{15}$$

Where ζ_{pilot}, ζ_{paging} and ζ_{sync} represent the power allocation percentages for the pilot, paging and synchronization channels, respectively.

ν denotes the voice activity factor.

H_{rf} represents the hand-off reduction factor which provides an estimate of the additional power needed for hand-off.

ζ_{tch} denotes the traffic channel power allocation percentage.

On the reverse link, the total interference level above the background noise level can be monitored as the reverse link rise value. For a given reverse link rise value (denoted as RL_rise, a value without units) under a corresponding number of active users (N_{active}) in the system, the reverse link capacity (N_{RL}) can be derived as

$$N_{RL} = \frac{N_{active}}{1-1/RL\_rise} \tag{16}$$

When the number of active users in the system is held constant, reverse link capacity decreases as the reported reverse link rise increases.

4. MICROCELL SIMULATION ANALYSIS

The CDMA system is an interference limited system. The required transmit powers at the mobile station and the base station must be large enough to satisfy the signal-to-noise ratio requirement, but small enough to avoid interference.

Simulations are carried out with a CDMA planning tool to analyze the system performance. Both the forward and reverse link powers are controlled at the minimum required power levels, with 1 dB standard deviation. Monte Carlo simulations are carried out on two tiers of hexagonal three-sectored cell sites, as illustrated in Fig. 3. The microcell is also three-sectored to obtain a consistent capacity comparison between the microcell and the macrocell. Mobile stations are placed randomly and uniformly over the service area of each sector. Therefore, when the cell site radius decreases, the traffic density in the cell increases. The statistics of capacity, hand-off percentage, signal and interference levels are obtained through simulation.

The information bit rate for the simulation is set at 14.4 kbps. The processing gain is 85.3. The orthogonality factor is 0.8. The pilot, paging, and synchronization channel power percentages are set at 15%, 12% and 1.5%, respectively. The macrocell base station power is set to 5 Watts. The microcell base station power is set between 0.42 Watts and 5 Watts.

The microcell and macrocell propagation models specified in Table I are used in the simulations. Flat terrain is assumed. The radius of the macrocell (denoted as R) is fixed at 3 miles. The radius of the microcell (denoted as r) varies from 0.3 to 0.6 miles. Five cases are studied and simulated as follows:

Case I: Regular cell sites without microcell. It is used as a baseline to determine macrocell capacity without embedded microcell.

Case II: A microcell is added at $2/3R$ from the macrocell site. The transmit power is equal to that of the macrocell. The microcell radius is about 0.6 miles.

Case III: The microcell power is reduced to one-third that of Case II. The cell radius is about 0.45 miles.

Case IV: The microcell power is reduced to one-quarter that of Case III. The cell radius is about 0.3 miles.

Case V: A microcell is added at $1/3R$ from the macrocell site. The transmit power is equal to that of the macrocell. The microcell radius is about 0.3 miles.

The preliminary results are summarized in Table II. The forward link capacity is defined as the maximum number of mobile stations that a system can support with a forward link failure rate of less than 5%. The reverse link capacity is calculated using (16). The voice activity factor used is 0.45 for the forward link and 0.40 for the reverse link.

The cell site capacity is defined as the smaller of forward and reverse link capacity. The microcell capacity gain g_μ is measured by the ratio of the microcell capacity to a regular cell site capacity. The combined network capacity gain g_s is defined as the ratio of the combined microcell and macrocell capacity to a regular cell site capacity.

Table II shows that the system capacity is limited by the forward link. The microcell capacity is 1.03 to 1.12 times the capacity of a regular cell site. The combined microcell and macrocell provide 2.00 to 2.11 times the capacity of a regular cell site. These results show that microcells could be a very efficient engine to relieve hot spot capacity requirement. The mobile stations are distributed continuously around microcell boundaries in this study. Therefore, the microcell capacity is lower than the case where a guard zone is assumed between the macrocell and microcell to isolate mutual interference [3].

Table II. The CDMA Hierarchical System Capacity

Case	Macrocell		Microcell		Microcell	Combined
	FL Capacity	RL Pole Capacity	FL Capacity	RL Pole Capacity	Capacity Gain ($g\mu$)	Capacity Gain (gs)
I	18	26.5				
II	17.46	25	18.5	31	1.03	2.00
III	17.83	27	19.0	30	1.06	2.05
IV	17.90	30	20.1	29	1.12	2.11
V	16.80	27	20.5	31	1.14	2.07

When the microcell is closer to the center of the macrocell (Case V), simulations show that the macrocell capacity decreases by one user, while the microcell capacity is greater than in Case IV.

The embedded microcell encounters less interference than in a uniform cluster, because the macrocell around the microcell has a much sparser mobile distribution. The microcell capacity is higher than the baseline. The microcell contributes additional interference to the macrocell, thus, reduces the macrocell capacity, especially when it is close to the macrocell site. However, the impact will be negligible when the microcell power is reduced and the microcell is placed farther away from the macrocell site, as in Case IV.

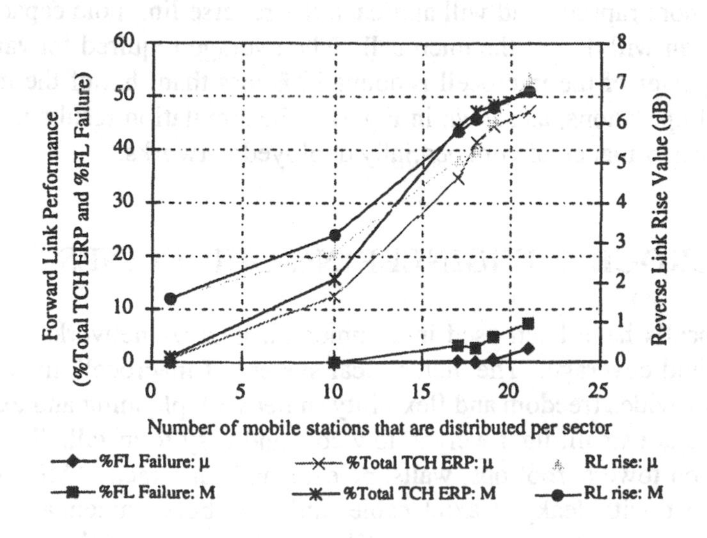

Fig. 5. Performance Evaluation of the CDMA System

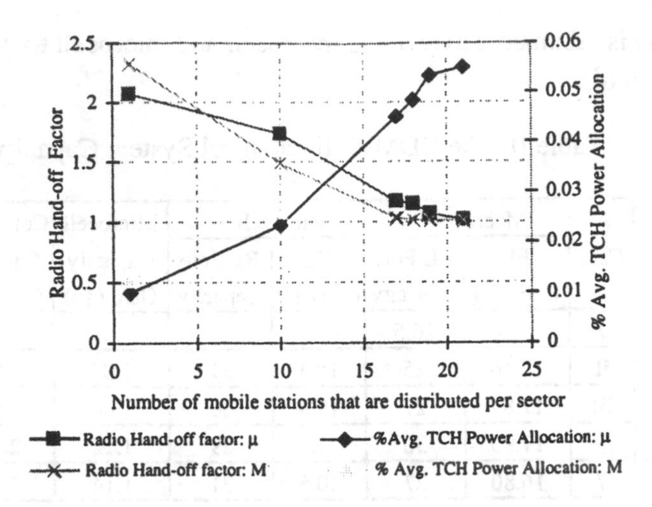

Fig. 6. Hand-off Factor and Average Forward Traffic Channel Power of the CDMA system

Microcell and macrocell performance is also analyzed in terms of RF reliability, soft hand-off factors, interference and power levels. RF reliability is measured by the percentage of mobile stations that failed to achieve the target voice quality in the intended coverage area. The macrocell RF reliability will degrade more seriously than that of the microcell as shown in Fig. 5. The radio hand-off factors (represents number of radio links needed for hand-off) of the microcell are about 11% higher than those for the macrocell at normal operations with about 10 mobile stations per sector, as shown in Fig. 6. The macrocell interference level will increase more rapidly, and will approach the reverse link pole capacity more quickly than will that of the microcell. The average required forward traffic channel power of the microcell is about 10% less than that of the macrocell at normal operations, as shown in Fig. 6. The simulation results match well with the performance of commercially deployed networks.

5. MICROCELL ENGINEERING GUIDELINES

Microcells have been used in commercial CDMA networks to enhance capacity and coverage. The hierarchical system of macrocell, microcell and picocell provides freedom and flexibility in network planning and expansion. Microcells are small, light weight, low cost and easy to install. They can be mounted on towers, rooftops, walls, poles, stands and racks. Microcells can interconnect with leaky coaxial cable and distributed antenna systems to expand indoor and tunnel coverage. These reduce cost and deployment time for network operators. The microcell system is fully integrated with the macrocell system for alarms, operation and maintenance management.

The CDMA network is a dynamic system. CDMA cell sites interact with each other closely and dynamically. When microcells are deployed in the middle of a commercial network, the interference and soft hand-off conditions of surrounding cell sites will change. Therefore, the system should be re-optimized to improve carried traffic capacity and performance. The pilot channel transmit power and antenna configuration of the surrounding cell sites may need to be adjusted. The network should be operated under minimal required transmit powers and minimal required soft hand-off areas. Sophisticated RF design tools will facilitate the optimization process by taking into account comprehensive power and interference management in the CDMA system.

A sufficient number of PN offsets must be reserved for microcell and picocell deployment. System PN offsets may also be re-assigned during system growth. Careful planning is required to account for some additional delays introduced by certain microcell applications, such as the time delay on optical fiber or other transmission media between remote RF units and the microcell main unit.

A typical picocell has 1 to 3 carrier-sectors, a microcell has 3 to 6 carrier-sectors, and a regular cell has 6 to 12 carrier-sectors. The RF transmit power of microcells and picocells is typically below 10 Watts for a single carrier-sector, while that of a regular cell site is above 10 Watts.

We envisage that future microcell base stations will include some functionality which, in a traditional network, is in the Base Station Controller (BSC). The hand-off and mobility management can be implemented in a streamlined manner in a distributed network architecture. The microcell base station will directly interface to an IP based core network to efficiently transmit packet data for internet applications. Microcells will then be able to support voice and data services with flexibility, modularity and expandability.

6. CONCLUSIONS

In this chapter, a practical CDMA microcell structure was systematically analyzed and simulated on both the forward link and reverse link. The results show that embedding the microcell in an existing CDMA network could be a very efficient way to improve hot-spot capacity and dead-spot coverage.

REFERENCES

[1] TIA/EIA/IS-95, "Mobile Station - Base Station Compatibility Standard for Dual-Mode Wideband Spread Spectrum Cellular System, " July 1993.

[2] Joseph Shapira, "Microcell Engineering in CDMA Cellular Networks", IEEE Transactions on Vehicular Technology, Vol. 43, No. 4, pp. 817-825, November 1994.

[3] Arthur Ross, "Overlay/Underlay Cells Efficiency," TR45.5 Contributions, TR45.5/92.10.

[4] Jung-Shyr Wu, Jen-Kung Chung and Yu-Chuan Yang, "Performance Improvement for a Hotspot Embedded in CDMA Systems," Proceedings of VTC, pp.944, May 1997.

[5] Howard Xia, Henry L. Bertoni, Leandro Maciel, Andrew Lindsay-Stewart, and Robert Rowe, "Radio Propagation Characteristics for Line-of-Sight Microcellular and Personal Communications," IEEE Trans. on Antennas and Propagation, Vol.41, No.10, pp. 1439-1447, October 1993.

[6] Jin Yang, Sung-Hyuk Shin and William C.Y. Lee, "Design Aspects and System Evaluations of IS-95 based CDMA System,"Proceedings of ICUPC, pp.381-385,October 1997.

[7] Allen Salmasi and Klein Gilhousen, "On the system design aspects of Code Division Multiple Access (CDMA) applied to digital cellular and personal communications networks," Proc. 41st IEEE VTC Conf., 1991, pp.57-62.

[8] Jin Yang, "Diversity Receiver Scheme and System Performance Evaluation for a CDMA System," IEEE Trans. on Communications, Vol. 47, No.2, pp. 272-280, February 1999.

Chapter 6

INTERMODULATION DISTORTION IN IS-95 CDMA HANDSET TRANSCEIVERS

STEVEN D. GRAY AND GIRIDHAR D. MANDYAM

Nokia Research Center, Irving, Texas

Abstract: Intermodulation distortion is a troublesome phenomenon that occurs in many wireless transceivers. The effects of intermodulation distortion often result in reduction of the dynamic range of wireless transceivers. This is of particular interest in IS-95 CDMA handsets, which must typically maintain linear behavior over a large dynamic range when compared to other public wireless systems. In this work, a theoretical framework for intermodulation distortion in handset receivers and transmitters is given. In addition, a method for detection of intermodulation distortion in IS-95 handset receivers is provided along with a means of mitigating such distortion.

1. INTRODUCTION

Intermodulation distortion is a phenomenon that occurs in wireless systems, and can be detrimental to wireless transceiver performance. This effect can impact both the receiver and transmitter in a wireless system. As a result, there exists much concern over reducing the degradation caused by intermodulation.

Intermodulation occurs as a result of the use of nonlinear components in typical wireless transceivers. When spurious interference is provided as input to nonlinear elements, this interference tends to appear in the output of these nonlinear elements as a linear interference term and several nonlinear interference terms. The nonlinear interference terms are often modeled as the weighted sum of powers of the input interference term, and therefore can be troublesome. Problems particularly arise when the nonlinear interference terms increase in power as a function of the input power levels to nonlinear components.

The effect of intermodulation is particularly troublesome in coherent-detection CDMA (code division multiple access systems). CDMA transceivers must normally operate over a large dynamic range, due to the need for feedback power control and pulse shaping. The most widely deployed CDMA system for public use today is the IS-95 system [1], which is implemented in many parts of North America, Korea, and Japan. The effect of intermodulation on this system is of particular interest for handset transceivers, as handsets' costs are impacted by linearity requirements in radio frequency (RF) components.

1.1 Intermodulation Theory

Ideal transceiver components will have a desired linear response to an infinite range of input signal levels. However, non ideal components may have a voltage output response given by

$$V_{out} = \sum_{n=0}^{\infty} a_n V_{in}^{\ n} \tag{1}$$

where V_{in} is the input signal voltage level, a_n is a scalar coefficient, and V_{out} is the output voltage level. In a communications system, this type of response results in undesired products. For instance, consider the case where the input signal is

$$V_{in} = A\cos(\omega_1 t) + B\cos(\omega_2 t) \tag{2}$$

In Equation (2), assume that the sinusoid at frequency ω_1 is the desired signal and the sinusoid at ω_2 is an undesired tone with sufficiently large spectral separation from the desired signal. The first two terms of the expansion of Equation (1) will be linear terms: a_0, and $a_1[A\cos(\omega_1 t)+B\cos(\omega_2 t)]$. a_0, or the DC term, is often considered negligible. a_1 is the device small-signal, or linear, gain. However, the second-order intermodulation product is

$$V_{out}^{(2)} = a_2\left\{\begin{array}{l}\frac{A^2}{2}(1+\cos(2\omega_1 t)) + \frac{B^2}{2}(1+\cos(2\omega_2 t)) + \\ AB[\cos((\omega_1-\omega_2)t) + \cos((\omega_1+\omega_2)t)]\end{array}\right\} \tag{3}$$

This second-order product contains a spectral component at ω_1-ω_2, which increases exponentially with the input signal by a power of 2. Moreover, the third-order term in the expansion is

$$V_{out}^{(3)} = a_3\left\{\begin{array}{l}(\frac{3A^3}{4} + \frac{3AB^2}{4})\cos(\omega_1 t) + \\ (\frac{3B^3}{4} + \frac{3A^2B}{4})\cos(\omega_2 t) + \\ \frac{A^3}{4}\cos(3\omega_1 t) + \frac{B^3}{4}\cos(3\omega_2 t) + \\ \frac{3AB^2}{4}[\cos((\omega_1 - 2\omega_2)t) + \cos((\omega_1 + 2\omega_2)t)] + \\ \frac{3A^2B}{4}[\cos((2\omega_1 - \omega_2)t) + \cos((2\omega_1 + \omega_2)t)]\end{array}\right\} \tag{4}$$

This third-order product is particularly troublesome in receiver design because the signals at frequencies $2\omega_1$-ω_2 and ω_1-$2\omega_2$ lie very close to the signal of interest in the spectral domain; in fact, many times this signal will fall on top of the desired signal bandwidth. Moreover, this term increases exponentially in magnitude by a power of 3 when compared to the desired linear term.

An intercept point is defined as the input signal power level at which an undesired higher-order output product is equal to the desired linear output signal. A device in transceiver design is often characterized by its intercept points. Of particular interest are the second and third-order intercept points, which correspond to the terms in Equations (3) and (4), respectively. Although the second-order intercept point is important, the third-order intercept point is considered more critical due to this product's proximity to the desired signal band. Nevertheless, the *n*th-order intercept point can be defined as

$$IPn = A + \frac{\Delta}{n-1} \tag{5}$$

where A is the input signal level in dBm and Δ is the difference in dB between the desired signal level and undesired product level. This concept is illustrated in *Figure 1*.

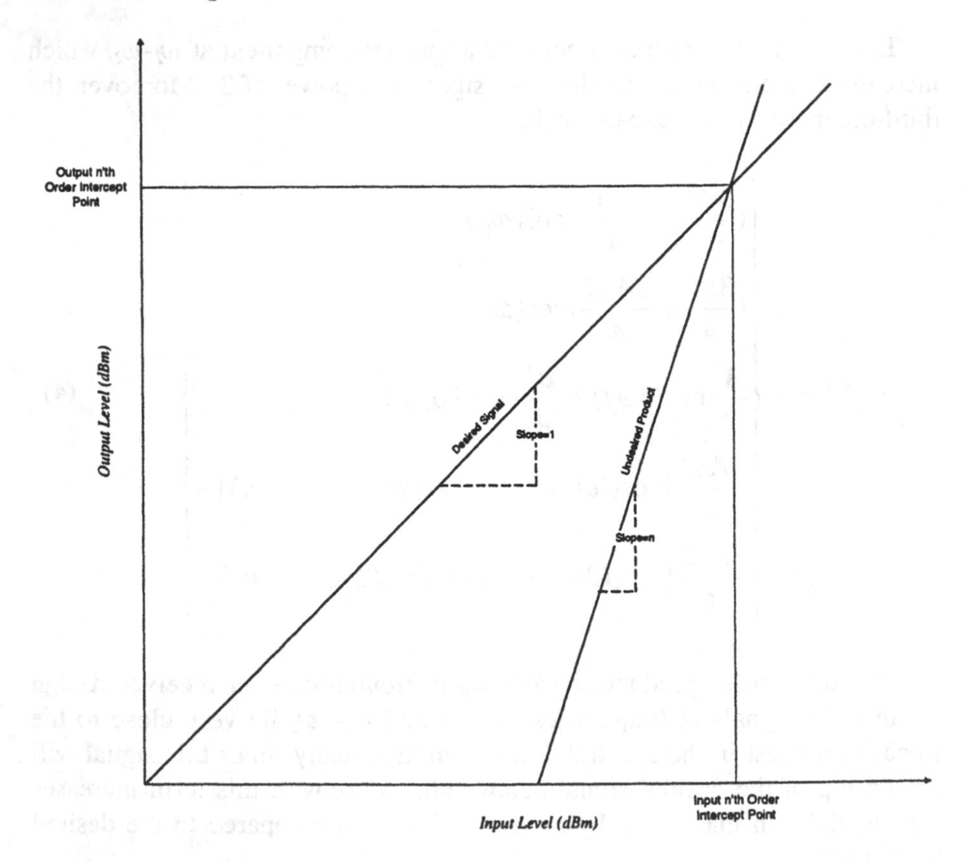

Figure 1. Intercept Point

Using the 2nd-order intercept point, a_2 can be found by setting A = Vip_2 and $B= 0$ in Equation (2), inferring that $V_{out}^{(2)} = V_{out}^{(1)}$ when:

$$a_2 = \frac{a_1}{Vip_2} \tag{6}$$

Similarly, a_3 may be found by setting $A = Vip_3$:

$$a_3 = \frac{4a_1}{3Vip_3^{\ 2}} \tag{7}$$

Since the third-order product increases cubicly compared to the desired linear product, it is necessary to have adequate filtering and gain control to reduce spurious products that may dominate at the input stages to nonlinear elements in the transceiver.

The intercept points of a system are useful for determining system performance; however, the intercept points are fictional in that the system or individual device will reach saturation well before the point when input signal levels are such that the intercept points are achieved. As a result, the intercept point is a useful measure for *small signal* analysis. *Large signal* analysis must take into account the compression point in addition to the intercept point. The 1-dB compression point for a system is defined as the input signal level at which the output voltage has decreased 1 dB with respect to the linear gain level. This value is closely related to the intercept point; using the previous analysis, we can see exactly how they are related. Returning to Equation (2), if the undesired term is excluded (i.e. $B = 0$), then the only contributions to the output are the linear term and the third-order term, then when $A = V_{IP}$, the ratio of linear gain to actual gain becomes

$$\frac{a_1 V_{cp}}{a_1 V_{cp} + a_3 \frac{3}{4} V_{cp}^{\ 3}} = 10^{\frac{1}{20}} \tag{8}$$

If the value for a_3 derived in Equation (7) is substituted Equation (8), then a closed-form solution for the relationship between the 1-dB compression point and the third-order intercept point may be derived:

$$10\log_{10}\left(\frac{V_{ip3}^{2}}{V_{cp}^{2}}\right)=10\log_{10}\left(\frac{1}{1-10^{\frac{-1}{20}}}\right) \quad (9)$$

This implies a 9.638 dB difference between 1-dB compression point and third-order intercept point. This relationship is a good rule-of-thumb, but cannot be guaranteed in practice.

2. RECEIVER INTERMODULATION

Receiver intermodulation usually results from the presence of interfering signals in the proximity of the receiver. These interfering signals induce intermodulation distortion in the nonlinear components present in a typical wireless receiver. For instance, in typical IS-95 CDMA systems operating in the "cellular band" (806-890 MHz) in North America, handsets often experience intermodulation distortion from AMPS (Advanced Mobile Phone Service) base station transmitters [2,3]. Since AMPS is a frequency-modulated narrowband system, the required carrier-to-interference ratios for these systems are much higher than IS-95 systems, and therefore intermodulation distortion in IS-95 receivers can degrade performance completely [4].

The sources of degradation can be more clearly evaluated when examining the receiver chain being used and isolating all nonlinear elements in this chain.

2.1 Wireless Receivers

A typical wireless terminal receiver for a full-duplex QPSK CDMA system (pictured in *Figure* 2) consists of several components; namely, the duplexor, the low-noise amplifier (LNA), the first mixer or IF mixer, a channel selection filter (often implemented as a surface acoustic wave (SAW) filter at IF), the automatic gain control (AGC) amplifier, the I-Q demodulator, the baseband antialiasing (AAF) filters, the analog-to-digital (ADC) converters, and the digital section.

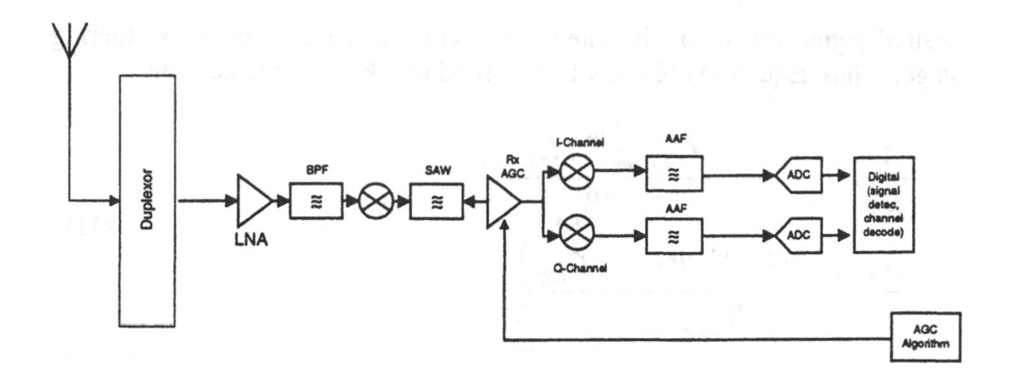

Figure 2. Receiver Chain

Partitioning of this functionality into three sections of the handset receiver is usually desirable. There three sections are the RF front-end, which encompasses all circuitry between the antenna and the antialiasing filters; the mixed-signal section, which includes the AAF filters and A/D converters; and finally the digital section.

The elements that will be subject to the greatest intermodulation distortion are the active components that operate at high frequencies. LNA's and other pre-amplifiers (e.g. AGC amplifiers), and mixers are examples of such devices. Passive elements, such as SAW filters, typically have a sufficiently large range of linearity so that they are often assumed to be infinitely linear.

In order to determine if a typical receiver chain can achieve the required intercept point, one needs to understand and evaluate this intercept point based on the individual components intercept points, passband gain, and selectivity. This can be found with the following formula:

$$\frac{1}{IP_3} = \frac{1}{IP_{3_{Duplexor}}} + \frac{G_{duplexor}}{IP_{3_{LNA}}} + \frac{G_{duplexor}G_{LNA}}{IP_{3_{SAW}}} + \ldots \quad (10)$$

Since passive elements such as a SAW filter display linear behavior over a large input range, their individual IP3's are assumed to be infinite. However, mixer and amplifier stages do not display such behavior. Another interesting aspect of the formula in Equation (10) is that, unlike noise figure, the second-to-last stage in the receiver chain has a large impact on system IP3. Finally, the cumulative gain in each stage must take into account selectivity, as the difference between the interference signal level and the desired signal level, for the interference can decrease after every filtering stage. Thus, Equation (10) may be modified to take this into account:

$$\frac{1}{IP_3} = \frac{1}{IP_{3_{Duplexor}}} + \frac{G_{duplexor}{}^{IM} / G_{duplexor}}{IP_{3_{LNA}}} + \frac{G_{duplexor}{}^{IM} G_{LNA}{}^{IM} / \left(G_{duplexor} G_{LNA}\right)}{IP_{3_{SAW}}} + \ldots \tag{11}$$

where $G_{(\,)}{}^{IM}$ is the gain at the intermodulation product frequency. Note that these equations only provide a first-pass analysis and that system IP3 calculation cannot be accurately determined without actual testing of a radio implementation.

2.2 Receiver Intermodulation Detection

One simple way of avoiding intermodulation interference is to prevent strong IP3 signals from occurring at the MS front-end prior to mixing. The proposed action is to switch out the amplifying element prior to mixing to avoid nonlinear distortion. The natural candidate for gain switching is the low noise amplifier, LNA, which is the first amplifier in the down-conversion process. The downside of gain switching the LNA is that it sets the noise figure for the RF portion of the receiver and eliminating it in the demodulation process increases the receiver noise figure.

The intermodulation products of most concern for an IS-95 mobile are due to AMPS carriers. Gain switching the LNA is a measure we wish to pursue only when the IP3 is severe. In most cases, the distance between the mobile and an AMPS base station will be large enough to allow the LNA to be in a high gain state. The problem addressed in this paper is one of detecting when the IP3 is large enough to degrade CDMA call quality. A positive detection will result in a low gain setting for the LNA. The signal used to assess when to switch the LNA is the baseband signal after A/D conversion.

To formulate a detector for AMPS IP3, a frequency domain stochastic model for the intermodulation interference and IS-95 signal is pursued. This model assumes that baseband in-phase and quadrature samples are converted to frequency domain samples by means of a discrete Fourier transform (DFT). The model presumes two distinct cases. The first is that only the CDMA signal is present in the desired spectrum. The second is that the

CDMA signal and intermodulation interference is present in the desired spectrum. The goal of this work is to develop a two hypothesis test to determine when intermodulation interference is present.

Interference and Signal Model

The detection problem is considered in the frequency domain where the discrete-time (sampled at a frequency f_c) complex baseband signal is given by $x[n]$.The DFT of $x[n]$ is

$$x_k = 1/L \sum_{n=0}^{L-1} x[n] \exp(-j2\pi nk/L) \tag{12}$$

Recalling that sums of Gaussian random variables produce Gaussian random variables, the DFT, x_k is distributed Gaussian under the assumption that it is Gaussian. In addition, the squared magnitude,

$$|x_k|^2 = \left[\mathrm{Re}(x_k)^2 + \mathrm{Im}(x_k)^2\right] \tag{13}$$

is distributed exponential.

Mathematical Model of Test

Considering the model for the case when intermodulation interference does and does not exist, two hypotheses are defined as follows:

$$\mathrm{H}_1: \; r_k[n] = a_k[n] 1_{K_1}(k) + b_k[n] 1_{K_2}(k) \qquad \text{Intermod Interference}$$

$$\mathrm{H}_2: \; r_k[n] = c_k[n] 1_K(k) \qquad \text{no Intermod Interference}$$

where

$$1_K(k) = \begin{cases} 1 & k \in K \\ 0 & \text{otherwise} \end{cases}, \tag{14}$$

$r_k[n]$ is the squared magnitude of the DSP/DFT processor output and the probability density functions (pdf's) for the model shown above are

$$f(a_k[n]) = 1/\lambda_{1,k} \exp[-a_k[n]/\lambda_{1,k}] \qquad a_k[n] \in [0,\infty)$$
$$f(b_k[n]) = 1/\lambda_{2,k} \exp[-b_k[n]/\lambda_{2,k}] \qquad b_k[n] \in [0,\infty)$$
$$f(c_k[n]) = 1/\gamma_k \exp[-c_k[n]/\gamma_k] \qquad c_k[n] \in [0,\infty).$$

The index k represents the DFT bin number and the index n indicates how the bin changes as a function of time. The pdf's shown above also indicate that a different pdf exists for each DFT bin. Under H_1, the set K_1 is defined to contain integer values corresponding to DFT bins with intermodulation interference and K_2 contains indices corresponding to DFT bins that do not have intermodulation interference. In addition, DFT bins that do not have intermodulation interference power under H_1 have statistics that differ from the DFT bins under H_2. This assumption is partially due to the adjustment in the dynamic range of the input signal prior to A/D conversion by an automatic gain control circuit when intermodulation interference is present.

Experimental Verification of the Model

To test the validity of the exponential model assumption, data was collected at the output of the DSP FFT processor during high intermodulation interference conditions and when the interference was considered insignificant. Based upon the visual classification of experimental data, a comparison is conducted between the empirical Cumulative Distribution Function (CDF) and the analytical CDF to determine how closely the data follows the assumption that the FFT bins are drawn from exponential random processes. *Figure 3* is a comparison between the empirical and analytical CDF (defined as $\Pr(C[n] < c_k[n])$ where $C[n]$ is a random variable) when the intermodulation interference was not significant as indicated under H_2. *Figure 4* is a similar comparison except for a DFT bin when the intermodulation interference was large.

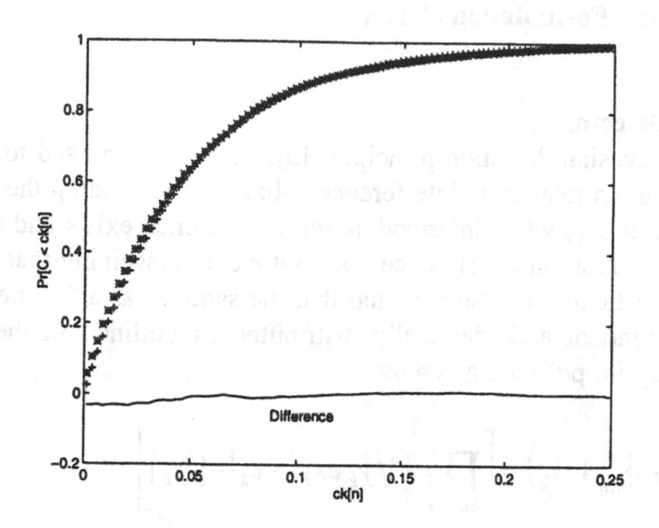

Figure 3. **Comparison between the Analytical and Empirical CDF for a Case with no Intermodulation Interference**

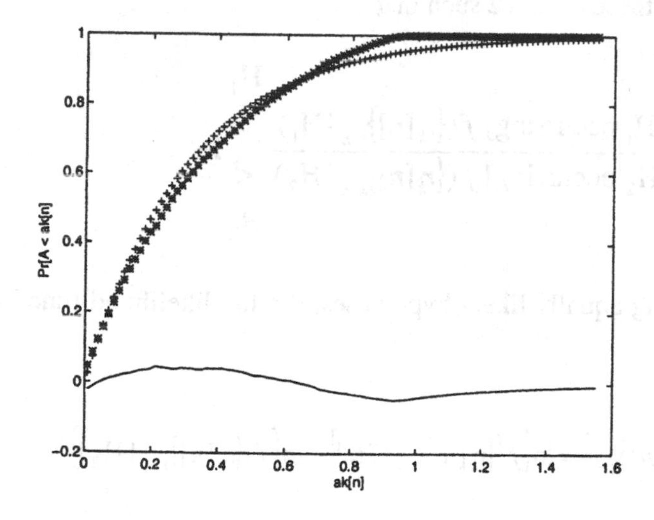

Figure 4. **Comparison between the Analytical and Empirical CDF for a Case with Intermodulation Interference**

Mathematical Formulation of Test

Bayesian Detector

Using Bayesian detection principles [5], the approach used to formulate the test for intermodulation interference is based upon finding the joint pdf's of $r_k[n]$ under H_1, when intermodulation interference exists and under H_2, when intermodulation interference does not exist. Assuming that each DFT bin, k is statistically independent and that the samples as a function of time, n are independent and identically distributed as outlined in the previous section, the joint pdf's are given by

$$f\left(\{r_k[n]\}_{k,n} \mid H_2\right) = \left[\prod_{k \in K} \prod_{n=1}^{N} 1/\gamma_k \exp[-c_k[n]/\gamma_k]\right] \tag{15}$$

where N is the number of time samples, for each DFT bin. The Bayesian test is constructed, assuming equal costs, by taking the ratio of the two joint pdf's under Hypotheses 1 and 2 such that

$$\frac{\Pr[\text{of } H_1 \text{ occurring}]\, f(\{r_k[n]\}_{k,n} \mid H_1)}{\Pr[\text{of } H_2 \text{ occurring}]\, f(\{r_k[n]\}_{k,n} \mid H_2)} \underset{H_2}{\overset{H_1}{\gtrless}} 1 \tag{16}$$

Assuming equally likely hypotheses, the log-likelihood function is given by

$$\Delta\left(\{r_k[n]\}_{k,n}\right) = \ln\left[f\left(\{r_k[n]\}_{k,n} \mid H_1\right)\right] - \ln\left[f\left(\{r_k[n]\}_{k,n} \mid H_2\right)\right] \tag{17}$$

where

$$\Delta\left(\{r_k[n]\}_{k,n}\right) = -N\sum_{k\in K_1}\ln(\lambda_{1,k}) - N\sum_{k\in K_2}\ln(\lambda_{2,k}) -$$
$$N\sum_{k\in K_1}\left(\bar{r}_k/\lambda_{1,k} - \bar{r}_k/\gamma_k\right) - N\sum_{k\in K_2}\left(\bar{r}_k/\lambda_{2,k} - \bar{r}_k/\gamma_k\right) + \tag{17}$$
$$N\sum_{k\in K_1}\ln(\gamma_k) + N\sum_{k\in K_2}\ln(\gamma_k)$$

and

$$\bar{r}_k = \frac{1}{N}\sum_{n=1}^{N} r_k[n] \tag{18}$$

The test for intermodulation interference becomes

$$\Delta(\{r_k[n]\}_{k,n}) \underset{H_2}{\overset{H_1}{\gtrless}} 0 \tag{19}$$

and the challenge is to simplify $\Delta(\{r_k[n]\}_{k,n})$ to an implementable expression.

The parameters γ_k, $\lambda_{1,k}$, and $\lambda_{2,k}$ are the means of the random processes for each DFT bin when intermodulation interference is present and when it is not. Due to baseband filtering, each bin is weighted differently making the mean value across DFT bins differ. However, from a practical perspective, these parameters are not known a priori and it is difficult to estimate separate means for each bin under the two hypotheses. As such, a simplification is made such that $\lambda_{1,k} = \lambda_1$, $\lambda_{2,k} = \lambda_2$, and $\gamma_k = \gamma$ for all k. Under this assumption, the log-likelihood test can be reduced to

$$\left(\frac{1}{\gamma} - \frac{1}{\lambda_1}\right)\sum_{k\in K_1}\bar{r}_k + \left(\frac{1}{\gamma} - \frac{1}{\lambda_2}\right)\sum_{k\in K_2}\bar{r}_k - \sum_{k\in K_1}\ln\left(\frac{\lambda_1}{\gamma}\right) - \sum_{k\in K_2}\ln\left(\frac{\gamma_2}{\gamma}\right) \underset{H_2}{\overset{H_1}{\gtrless}} 0 \tag{20}$$

Ideally, the above expression suggests that if the position of the intermodulation interference were a fixed and known quantity under H_1, then the test can be implemented by: (1) averaging in time each DFT bin, (2) dividing the DFT bins according to the sets K_1 and K_2 and (3) comparing the sums under K_1 and K_2 to a threshold. Unfortunately, the exact integer values within K_1 and K_2 are typically not known. In other words, the exact DFT bin position of the intermodulation interference is not known under H_1 prior to receiving the DFT output. In addition, λ_1, λ_2 and γ may not be known.

To simplify the test, we divide the test statistic shown above by

$$\left(\frac{1}{\gamma}-\frac{1}{\lambda_1}\right)\sum_{k\in K_2}\bar{r}_k \tag{21}$$

This allows the test to be written as

$$\frac{\frac{1}{k_1}\sum_{k\in K_1}\bar{r}_k}{\frac{1}{k_2}\sum_{k\in K_2}\bar{r}_k} \underset{H_2}{\overset{H_1}{\gtrless}} \frac{\ln\left(\frac{\lambda_1}{\gamma}\right)+\frac{k_2}{k_1}\ln\left(\frac{\lambda_2}{\gamma}\right)}{\left(\frac{1}{\gamma}-\frac{1}{\lambda_1}\right)\frac{1}{k_2}\sum_{k\in K_2}\bar{r}_k[n]}-\frac{\left(\frac{1}{\gamma}-\frac{1}{\lambda_2}\right)k_2}{\left(\frac{1}{\gamma}-\frac{1}{\lambda_1}\right)k_1} \tag{22}$$

where $k_1 = \dim(K_1)$ and $k_2 = \dim(K_2)$. Further simplification can be made by assuming that $\lambda_1 > \gamma >>> \lambda_2$ and that

$$\frac{\ln\left(\frac{\lambda_1}{\gamma}\right)+\frac{k_2}{k_1}\ln\left(\frac{\lambda_2}{\gamma}\right)}{\left(\frac{1}{\gamma}-\frac{1}{\lambda_1}\right)\frac{1}{k_2}\sum_{k\in K_2}\bar{r}_k[n]} << \frac{\left(\frac{1}{\gamma}-\frac{1}{\lambda_2}\right)k_2}{\left(\frac{1}{\gamma}-\frac{1}{\lambda_1}\right)k_1} \tag{23}$$

Using a law of large number assumption for $r_k[n]$, the above relationship is reasonable because the AGC attempts to keep the dynamic range of the A/D matched to the input signal causing a substantial difference in the apparent CDMA power per DFT bin when the interference is present and when it is not.

Using the approximations of the previous paragraph, the detector is as follows:

1) Compute,

$$\hat{S} = \frac{\sum_{k \in K_1} \bar{r}_k}{\sum_{k \in K_2} \bar{r}_k}$$

2) Compute,

$$t = \frac{\left(\frac{1}{\gamma} - \frac{1}{\lambda_2}\right)}{\left(\frac{1}{\lambda_1} - \frac{1}{\gamma}\right)}$$

3) If $\hat{S} > t$ intermodulation interference is present.
   If $\hat{S} < t$ intermodulation interference is not present

False Alarm Rate and Power

To complete the mathematical development, two measures of the test presented in the previous section are formulated. The first measure defined as

$$\delta = \Pr[\hat{S} > t \mid H_1] = \int_t^{\infty} f_S(\hat{s} \mid H_1)\, ds \tag{24}$$

is called the power in the test and represents the probability of deciding intermodulation interference exists when intermodulation interference does exist. The second measure defined as

$$\zeta = \Pr[\hat{S} > t \mid H_2] = \int_t^{\infty} f_S(\hat{s} \mid H_2)\, ds \tag{25}$$

is called the false alarm rate and represents the probability of deciding that intermodulation interference exists when none is present. To calculate δ and ζ, the probability density functions for these two measures must be obtained.

The expression for the pdf of the test statistic under H_1 is

$$f_S(\hat{s}\,|\,\mathrm{H}_1) = \left[\frac{\hat{s}^{n_1-1}}{B(n_1,n_2)}\right]\left[\frac{\lambda_1^{-n_1}\lambda_2^{-n_2}}{\left(\frac{\hat{s}}{\lambda_1}+\frac{1}{\lambda_2}\right)^{n_1+n_2}}\right]1_{(0,\infty)}(\hat{s}) \tag{26}$$

The above distribution is sometimes called a Beta distribution of the second kind. To find the pdf of the test statistic under H_2, substitute $\lambda_1 = \lambda_2 = \gamma$, in the above pdf conditioned on H_1. Further development of false alarm rate and power are presented in the next section with the experimental data.

Experimental Detection Results

The data used to assess the detection models was collected under heavy loading of AMPS base stations in the Los Angeles area. In many cases, the DSP FFT processor outputs were acquired as the distance between the CDMA mobile and a CDMA base station was increasing and the distance between the AMPS base station and the CDMA mobile was decreasing. Data was acquired until the call was dropped.

A major aspect of implementing the detector is in determining the values for the parameters λ_1, λ_2, γ, K_1 and K_2. The technique used for the Los Angeles data is to form sample estimates for λ_1, λ_2, and γ when intermodulation interference is present and when it is not. Given the fact that

$$\begin{aligned} E(a_k[n]) &= \lambda_1 \\ E(b_k[n]) &= \lambda_2 \\ E(c_k[n]) &= \gamma, \end{aligned} \tag{27}$$

a simple average across time and DFT bins is used to estimate each parameter. In addition, the number of DFT bins used in the test for the data presented in this section is such that $k_1 = k_2 = 4$. However, this is implemented by rank ordering $\bar{r}_k$ from the lowest bin to the highest bin and letting $K_1 = \{61,62,63,64\}$ and $K_2 = \{1,2,3,4\}$. In this case, bins $\{5,\ldots,60\}$ are discarded. The purpose for discarding midrange bins is to increase the separation between possible interference bins and non-interference bins.

From a numerical perspective, the detector was implemented by first normalizing each DFT record such that

$$\sum_{k=1}^{64} r_k[n] = 1 \tag{28}$$

Estimating the means of the exponential variables from known data, $\lambda_1 = 0.35$, $\lambda_2 = 0.005$ and $\gamma = 0.15$, yields a threshold of $t = 51$. The depth of the time average for computing $\bar{r}_k$ was set at five samples, $N = 5$. This corresponds to a decision once every 0.1 seconds.

Figure 5 and *Figure 6* are plots of the test statistic, $\hat{S}$, as a function time by starting far away from an AMPS base station and driving progressively closer. In *Figure 5*, the IP3 algorithm is in state H_2 (no intermodulation interference) until approximately 80 seconds into the test when $\hat{S} > t$ ($t = 51$). For *Figure 6* this occurs at 27 seconds in to the test. A characteristic of *Figure 5* and *Figure 6* are that the means of the test statistic under H_2 are approximately the same.

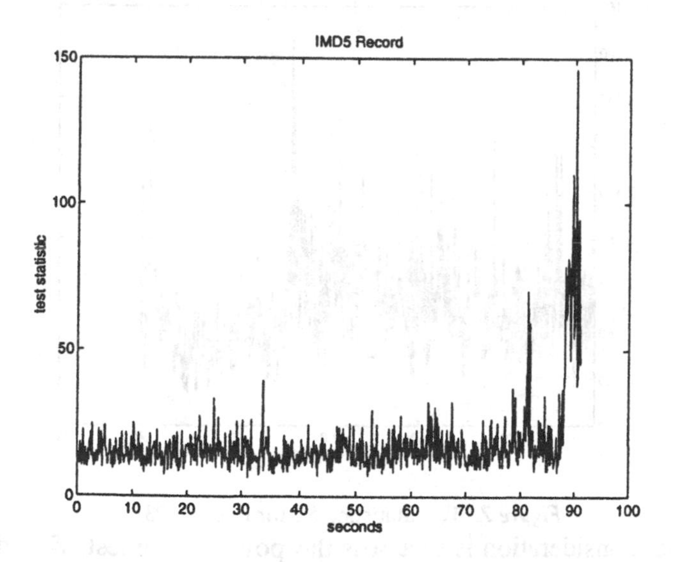

Figure 5. Test Statistic, S, for Record IP35

The last case considered in this section is when the data acquisition system was stationary at a position close to the AMPS base station. The test statistic is plotted for this case in *Figure 7*. The intermodulation interference is high and the IP3 algorithm immediately detects the interference.

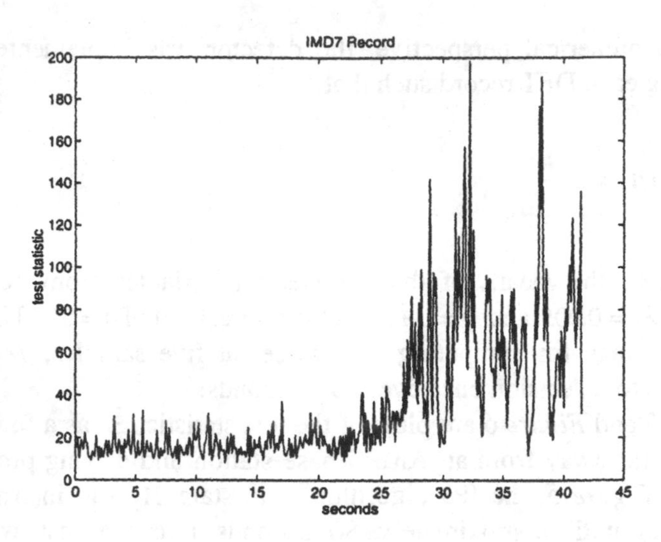

Figure 6. Test Statistic, S, for Record IP37

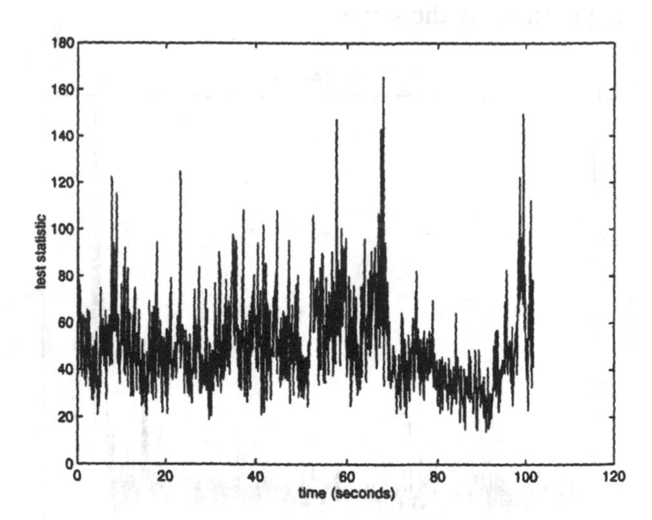

Figure 7. Test Statistic, S, for Record IP312

The final consideration is to assess the power in the test, δ and the false alarm rate, ζ. This is best understood by plotting $\Pr[S > t]$ under H_1 and H_2. Referring to *Figure 8,* the two measures are plotted based upon the pdf of the test statistic derived in the previous section and estimates listed previously for λ_1, λ_2, and γ. The "Under H_1" label corresponds to the power and the "Under H_2" label corresponds to the false alarm rate. For the data considered, a threshold, $t \approx 40$ yields a false alarm rate, $\zeta \approx 0$ and a power, $\delta = 0.96$. At the Bayesian setting of $t = 51$, the false alarm rate is near zero and the power, $\delta \approx 0.85$.

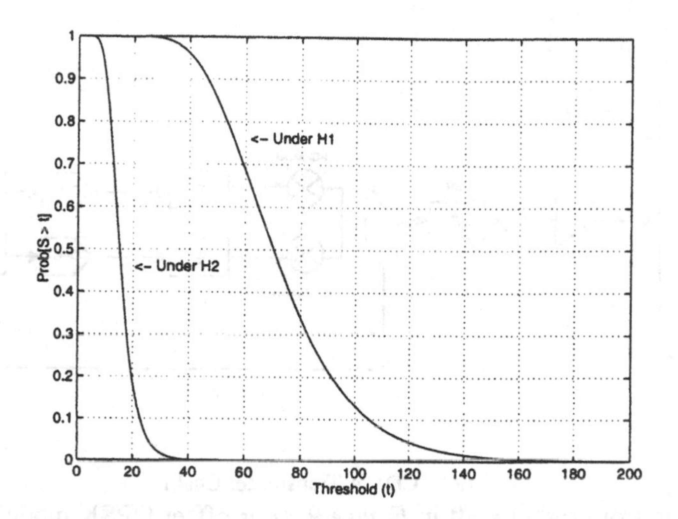

Figure 8. False Alarm Rate and Power in the Test for AMPS Test Data

The results of *Figure 5* thru *Figure 8* suggest that the detection of intermodulation interference is possible using a relatively simple algorithm. Switching the LNA from a high gain mode to a low gain mode when IP3 is present minimizes changes to existing baseband processing on most IS-95 CDMA handset while reducing the effects of interference. The detection results presented in this chapter and implementation results of the IP3 algorithm on a DSP processor demonstrate the performance and implementability of the IP3 detection algorithm. Furthermore, field testing has also demonstrated that switching the LNA in connection with detecting IP3 is successful in eliminating the problem of dropping calls when a CDMA handset is very close to an AMPS base station and far from a CDMA base station.

3. TRANSMITTER INTERMODULATION

Transmitter-induced intermodulation can be just as troubling as receiver intermodulation, due to the fact that the required modulation becomes compromised as a result of internal intermodulation products. This effect in turn makes it difficult to manufacture radio which meet mandated electromagnetic compatibility requirements in addition to maintaining required modulation accuracy. An overview of the CDMA transmitter chain is shown below:

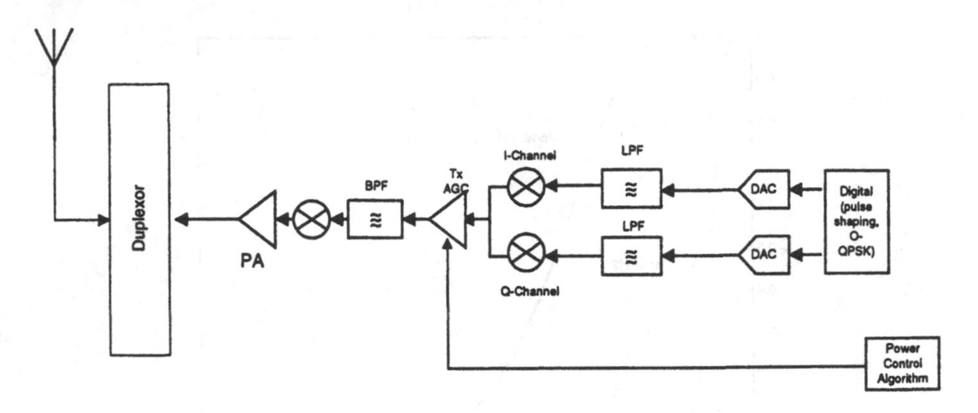

Figure 9. CDMA Transmitter Chain

Working from right to left in *Figure 9*, after offset-QPSK modulation and pulse shaping, the I and Q signals respectively drive two fixed-resolution digital-to-analog converters (DAC's). These DAC's in turn are filtered to reject sampling images and to meet electromagnetic compatibility requirements. These signals are now converted to an IF-frequency by modulation with orthogonal waveforms (IQ-modulation) and passed into a transmit AGC (Tx AGC) amplifier. This signal in turn is filtered for rejection of spurious products and upconverted to the desired carrier frequency. The signal is passed into a power amplifier PA and then through the duplexor to the antenna.

The Tx AGC amplifier along with the PA provide the necessary dynamic range for CDMA power control to operate. This requires a suitable range of linearity (at least 80 dB, according to the minimum performance specifications for IS-95 terminals IS-98-A [6]). These two amplifiers are also subject to intermodulation distortion; however, the sources of the interference are internal, and are usually provided by the upconversion stages present in the transmitter.

3.1 Sources of Transmitter Internal Interference

There are two primary sources of interference that can result in intermodulation in the transmitter chain pictured in *Figure 9*: the IQ-modulation and the upconverter.

IQ-Modulator

The IQ-modulator takes baseband input from both the I and Q paths and converts it to an IF frequency by modulating each respective path by mutually orthogonal waveforms (usually phase-synchronous since and cosine waves). In addition to the desired waveform, the IQ-modulator also provides undesired interference terms which can result in intermodulation within the Tx AGC amplifier.

The undesired interference terms can be traced back to gain and phase mismatch between the IQ and Q paths. The mismatch in gain and phase between the I and Q paths manifests itself in the output of the IQ mixer as two effects: sideband leakage and carrier leakage. In order to understand these effects, a mathematical model for the IQ-mixer may be used. A typical IQ-mixer is shown below:

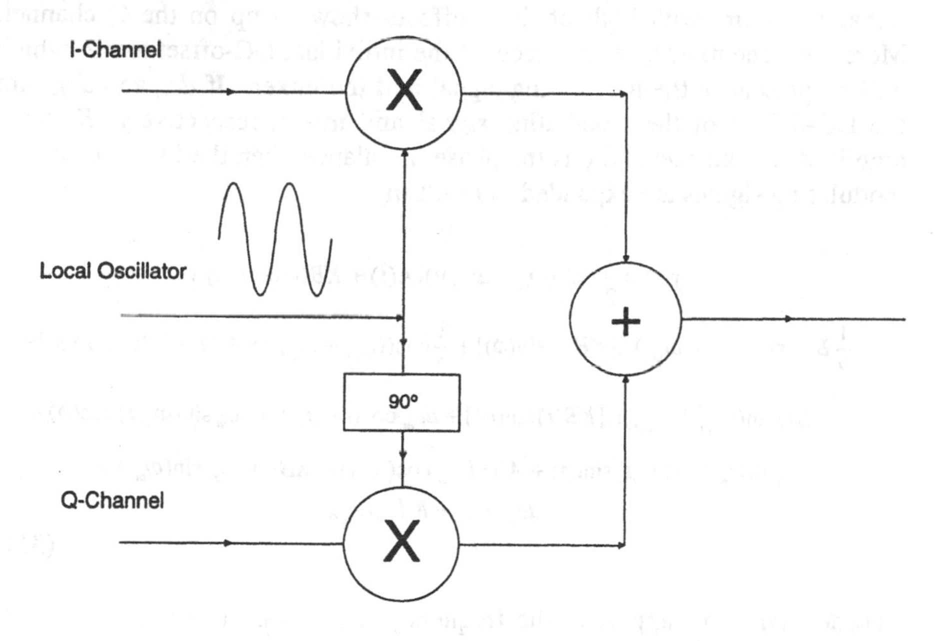

Figure 10. IQ-Mixer

Typically, the local oscillator frequency is referred to as ω_{LO}. The 90-degree phase shift results in orthogonality between the modulating signals; this results in quadrature amplitude modulation (QAM). The signals coming in on the I and Q channels may be represented as

$$i(t) = A(t)\cos(\omega_m t)$$
$$q(t) = B(t)\sin(\omega_m t) \quad (29)$$

where $A(t)$ and $B(t)$ are the I and Q information signals and ω_m is the modulation frequency. The ideal output of the IQ-mixer would be [7]

$$y(t) = i(t)\cos(\omega_{LO}t) + q(t)\sin(\omega_{LO}t) \quad (30)$$

However, due to amplitude and phase imbalances in the modulating signals and amplitude and phase imbalances within the mixer, this ideal output does not occur in reality. For the sake of this analysis, it is assumed that both the amplitude and phase imbalances of the modulating signal and the mixer can be combined into one amplitude imbalance and one phase imbalance term, with both of these effects showing up on the Q channel. Moreover, one must take into account the individual DC-offset terms, which will be present in the modulating signal and the mixer. If dc_m and dc_{LO} are the DC-offsets of the modulating signal and mixer, respectively; K is the amplitude imbalance; and ϕ is the phase imbalance, then the I and Q channel modulating signals are expanded to result in

$$\begin{aligned} y(t) = \frac{1}{2}\cos((\omega_m - \omega_{LO})t)[A(t) + KB(t)\cos(\phi)] + \\ \frac{1}{2}B(t)\sin((\omega_m - \omega_{LO})t)[KB(t)\sin(\phi)] + \frac{1}{2}\cos((\omega_m + \omega_{LO})t)[A(t) - KB(t)\cos(\phi)] + \\ \frac{1}{2}B(t)\sin((\omega_m + \omega_{LO})t)[KB(t)\sin(\phi)] + dc_m\cos(\omega_{LO}t) + Kdc_m\sin(\omega_{LO}t)\cos(\phi) + \\ Kdc_m\cos(\omega_{LO}t)\sin(\phi) + A(t)dc_{LO}\cos(\omega_m t) + KB(t)dc_{LO}\sin(\omega_m t) + \\ dc_m dc_{LO} + Kdc_m dc_{LO} \end{aligned} \quad (31)$$

The desired terms appear at the frequency ω_m - ω_{LO}. The other terms are undesired and can be problematic. Usually, the DC terms and the terms corresponding to the modulation frequency ω_m are easy to filter if the desired term is sufficiently high in frequency. However, two extremely problematic products are the sideband products, corresponding to the frequency ω_m + ω_{LO}, and the carrier products, corresponding to the frequency ω_{LO}. For simplification purposes, let as assume the $A(t)$ and $B(t)$ are power normalized such that their power is always equal to 1. The ratio of the power of the sideband products to the desired products is known as *sideband suppression*, and can be evaluated as

$$Sideband\ Suppression\,(dB) = 10\log_{10}\left(\frac{1-2K\cos(\phi)+K^2}{1+2K\cos(\phi)+K^2}\right) \quad (32)$$

The ratio of the power of the carrier frequency, or LO frequency, products to the desired products is known as the carrier suppression, and may be evaluated as

$$Carrier\ Suppression\,(dB) = 10\log_{10}\left(\frac{dc_m^{\ 2}+2Kdc_m dc_{LO}\sin(\phi)+K^2 dc_m^{\ 2}}{\frac{1}{4}\left(1+2K\cos(\phi)+K^2\right)}\right) \quad (33)$$

In a transmit IQ-mixer, the carrier terms and sideband terms tend to fall extremely close, if not on top of, the desired signal in the spectral domain. This is due to the modulating frequency being at or near baseband. Therefore, one must take care in baseband design to allow sufficient margin for the distortion an IQ-mixer may provide. However, one may alleviate the problems of carrier and sideband leakage through DC-offset correction, amplitude balancing, and phase balancing in the input stages to the IQ-modulator.

Upconverter

The upconverter serves to translate a signal at an IF frequency to the final carrier frequency. The interference that it produces will in turn drive the PA and can result in unwanted products in the transmitted signal. In this section, an image-rejecting upconverter is analyzed; this type of upconverter has inherent rejection of the undesired sideband. The upconverter is a high-frequency device, whose modeling is similar to the IQ-modulator. A common method to implement upconversion by means of the image-reject mixer is shown in *Figure 11*.

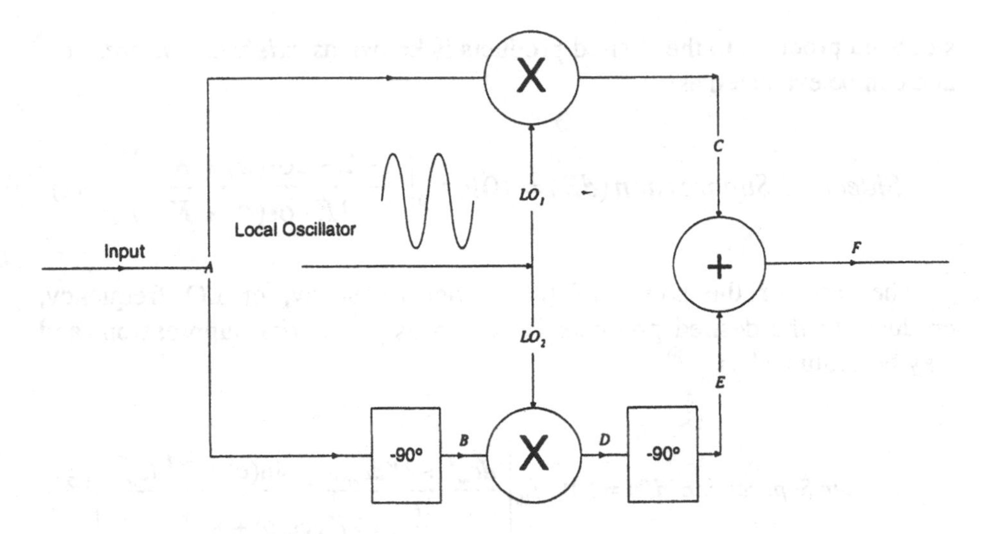

Figure 11. Image-Reject Mixer

This mixer has the advantage of suppressing the undesired sideband. This can be seen by modeling the response of this mixer to a sinusoid. Assuming that an input signal at frequency ω_{IF} is to be upconverted to a frequency at $\omega_{IF} + \omega_{LO}$, then two input sinusoids are required at frequencies ω_{IF} and ω_{LO}. Referring back to *Figure 11*, then the input signal at point *A* is given by

$$A = I\cos(2\pi\omega_{IF}t) + dc_{IF} \tag{34}$$

where *I* is the amplitude of the input, and dc_{IF} is a DC-offset. This signal is passed through a 90-degree hybrid coupler [8] to produce the signal at point *B*:

$$B = I\cos\left(2\pi\omega_{IF}t + \frac{2\pi}{4(\omega_{LO} - \omega_{IF})}\right) + dc_{IF} \tag{35}$$

The signal at points *B* and *A* are now to be modulated with a sinusoid at frequency ω_{LO} to produce the signals at points *C* and *D*. However, much like the IQ-mixer, this local oscillator signal will be subject to gain and phase imbalances. Thus the signals at LO_1 and LO_2 may be represented as

$$\begin{aligned} LO_1 &= Lc\cos(2\pi\omega_{LO}t + \theta) + dc_{LO} \\ LO_2 &= L\cos(2\pi\omega_{LO}t) + dc_{LO} \end{aligned} \tag{36}$$

where L is the nominal amplitude, c is the gain mismatch scaling constant, t is the phase mismatch, and dc_{LO} is the DC-offset of the LO signal. Thus the signal at point C can be represented as

$$C = \frac{\cos(\theta)ILc}{2}\cos(2\pi(\omega_{LO}+\omega_{IF})t) - \frac{\sin(\theta)ILc}{2}\sin(2\pi(\omega_{LO}+\omega_{IF})t) + \frac{\cos(\theta)ILc}{2}\cos(2\pi(\omega_{LO}-\omega_{IF})t) - \frac{\sin(\theta)ILc}{2}\sin(2\pi(\omega_{LO}-\omega_{IF})t) + Lcdc_{IF}\cos(\theta)\cos(2\pi\omega_{LO}t) - Lcdc_{IF}\sin(\theta)\sin(2\pi\omega_{LO}t) + Idc_{LO}\cos(2\pi\omega_{IF}t) + dc_{IF}dc_{LO} \quad (37)$$

The signal at point D results from the modulation of the signal at point A passed through the 90-degree hybrid. This signal is:

$$D = \frac{IL}{2}\cos\left(2\pi(\omega_{LO}-\omega_{IF})t - \frac{2\pi}{4(\omega_{LO}-\omega_{IF})}\right) + \frac{IL}{2}\cos\left(2\pi(\omega_{LO}+\omega_{IF})t + \frac{2\pi}{4(\omega_{LO}-\omega_{IF})}\right) + Ldc_{IF}\cos(2\pi\omega_{LO}t) + Idc_{LO}\cos\left(\frac{2\pi}{4(\omega_{LO}-\omega_{IF})}\right)\cos(2\pi\omega_{IF}t) - Idc_{LO}\sin\left(\frac{2\pi}{4(\omega_{LO}-\omega_{IF})}\right)\sin(2\pi\omega_{IF}t) + dc_{IF}dc_{LO} \quad (38)$$

The signal at point D is now passed through another 90-degree hybrid to yield

$$E = \frac{IL}{2}\cos\left(2\pi(\omega_{LO}-\omega_{IF})t - \frac{\pi}{(\omega_{LO}-\omega_{IF})}\right) + \frac{IL}{2}\cos(2\pi(\omega_{LO}+\omega_{IF})t) + Ldc_{IF}\cos\left(2\pi\omega_{LO}t - \frac{2\pi}{4(\omega_{LO}-\omega_{IF})}\right) + Idc_{LO}\cos\left(2\pi\omega_{IF}t - \frac{2\pi}{4(\omega_{LO}-\omega_{IF})}\right) + dc_{IF}dc_{LO} \quad (39)$$

Referring back to Equation (37), one can see the terms corresponding to the undesired sideband at ω_{LO}-ω_{IF} are eliminated under ideal conditions when the signals at points *E* and *C* are summed together to yield *F*. This is due to the fact that when the phase mismatch θ is 0 degrees, the term in *C* corresponding to $sin(2\pi(\omega_{LO}-\omega_{IF})t)$ disappears and the term in *C* corresponding to $cos(2\pi(\omega_{LO}-\omega_{IF})t)$ is 180 degrees out of phase with the corresponding $cos(2\pi(\omega_{LO}-\omega_{IF})t)$ term in *E*.

Of course, in reality the signal at *F* will have many undesired products, as indicated by the equations above. The four primary products present at the output of the image-reject mixer are the desired upper sideband, the lower sideband leakage, the IF leakage, and the LO leakage. The desired signal power may be expressed as

$$Desired = \frac{1}{2}\left(\frac{\cos(\theta)ILc}{2}\right)^2 + \frac{1}{2}\left(\frac{IL}{2}\right)^2 \tag{40}$$

The lower sideband, LO and IF powers may be expressed as

$$Sideband = \frac{1}{2}\left(\frac{\cos(\theta)ILc}{2} - \frac{IL}{2}\right)^2 + \frac{1}{2}\left(\frac{\sin(\theta)ILc}{2}\right)^2 \tag{41}$$

$$IF = \frac{1}{2}(L^2c^2dc_{IF}{}^2 + I^2dc_{LO}{}^2) \tag{42}$$

$$LO = \frac{1}{2}(I^2dc_{LO}{}^2 + L^2dc_{IF}{}^2) \tag{43}$$

The DC terms at point *F* are usually filtered by a DC-block and are not generally considered of consequence. However, the lower sideband term, the LO term, and the IF term may not be very easy to filter out. Of particular interest are the LO and IF isolation values, defined as

$$LO\ Isolation\ (dB) = 10\log_{10}\left(\frac{Desired}{LO}\right) \tag{44}$$

$$IF\ Isolation\ (dB) = 10\log_{10}\left(\frac{Desired}{IF}\right) \tag{45}$$

The analysis presented above assumed the gain and phase mismatch to be manifest in the LO signal to simplify analysis. However 90-degree transformer hybrids usually suffer from gain imbalances and phase mismatch. If one chooses another topology to accomplish the phase shifting, gain and phase matching still will be critical for performance. It should also be noted that the summing amplifier needed for summing C and E together to produce F may behave non-linearly over much of the upconverter's dynamic range. However, advanced circuit topologies may be able to alleviate this problem.

3.2 Transmitter Components subject to Intermodulation

The Tx AGC amplifier and PA are nonlinear elements in a CDMA system which must maintain a suitably large range of linearity.

Transmit AGC Amplifier

The transmit AGC amplifier is used primarily for power controlling the handset. These amplifiers typically support at least 80 dB of dynamic range, as per the requirements of IS-98A. Typically, the third-order polynomial model given in Section 1 is appropriate for modeling such amplifiers. However, an AGC amplifier is variable gain, meaning that its intercept point changes with each gain setting. The worst-case intercept point occurs when the amplifier is at maximum gain; therefore, it makes sense to verify operation at the maximum gain setting of the AGC amplifier. For instance, the part described in [10] provides a -18 dBm input third-order intercept point at 35 dB of gain setting. At gain settings greater than 35 dB, the intercept point reduces quickly to -27 dBm at 45 dB of gain setting. The part described in [10] provides provides -26 dBm at 39 dB maximum gain setting.

Power Amplifier

CDMA PA's typically must operate over a large dynamic range. As given in [11], the input and output voltages for the power amplifier may be given as

$$v_i(t) = \mathrm{Re}\{\rho e^{j2\pi f_c t}\}$$
$$v_o(t) = \mathrm{Re}\{A(|\rho|)e^{j2\pi f_c t + j\theta(|\rho|)}\} \qquad (46)$$

where f_c is the carrier frequency, ρ is the input complex magnitude, and $A(|\ |)$ and $\theta(|\ |)$ represent the gain and phase characteristics of the PA. The phase distortion term is to represent the AM to PM effect seen in many nonlinear devices. These functions can normally be derived from a single-tone test, where the effects of PA distortion on a singular frequency are captured. In [11], the authors demonstrate that by using this model, they were able to effectively simulate their two-stage gallium-arsenide MESFET amplifier and demonstrate the device's compliance with emission requirements for PCS CDMA.

Due to gain and phase distortion, spectral re-growth results in at the output of the PA. This is troublesome due to the need to meet specific electromagnetic compatibility requirements. Moreover, input stages to the PA, which provide spurious products will contribute difficulty in meeting the required spectral emissions mask.

4. CONCLUSIONS

Intermodulation distortion in IS-95 handset transceivers is particularly troublesome for both reception and transmission. However, if one can isolate the source of the interference resulting in intermodulation, one can compensate for this in either the receive or transmit paths. For the receiver, accurate detection of the presence of intermodulation is important. Once this is achieved, then appropriate action may be taken to ensure that intermodulation products do not capture the receiver. For the transmitter, intermodulation compensation may be accomplished by IQ-balancing and DC-offset compensation.

REFERENCES

[1] *TIA/EIA/IS-95-A: Mobile Station-Base Station Compatibility Standard for Dual-Mode Wideband Spread Spectrum Cellular System.* The Telecommunication Industry Association.

[2] Hamied, Khalid and Gerald Labedz. "AMPS Cell Transmitter Interference to CDMA Mobile Receiver." *IEEE Vehicular Technology Conference.* May, 1996. pp. 1467-1471.
[3] Joyce, Timothy. "Field Testing of QCP800 Phones in High Analog Interference Conditions." Ameritech Report. March, 1996.
[4] Shen-De, Lin, et.al. "Impact of CDMA Mobile Receiver Intermodulation on Cellular 8 Kbps System Performance." Lucent Technologies Report. February, 1996.
[5] Kazakos, D. and P. Papantoni-Kazakos. *Detection and Estimation.* New York: Computer Science Press, 1990.
[6] *TIA/EIA/IS-98-A: Recommended Minimum Performance Standards for Dual-Mode Wideband Spread Spectrum Cellular Mobile Stations.* The Telecommunication Industry Association.
[7] Umstattd, Ruth. "Operating and Evaluating Quadrature Modulators for Personal Communication Systems." *National Semiconductor Application Note 899.* October, 1993.
[8] Maas, Stephen A. *Microwave Mixers.* Norwood, MA: Artech House, 1986.
[9] Qualcomm, Inc. *Automatic Gain Control Amplifier Data Book.* July, 1997.
[10] RF Micro Devices. RF9909: *CDMA/FM Transmit AGC Amplifier.* Preliminary specification.
[11] Struble, Wayne, Finbarr McGrath, Kevin Harrington, and Pierce Nagle. "Understanding Linearity in Wireless Communication Amplifiers." *IEEE Journal of Solid State Circuits.* Vol. 32. No. 9. September, 1997. pp. 1310-1318.

PART III

DEPLOYMENT OF TDMA BASED NETWORKS

Chapter 7

HIERARCHICAL TDMA CELLULAR NETWORK WITH DISTRIBUTED COVERAGE FOR HIGH TRAFFIC CAPACITY

JÉRÔME BROUET*, VINOD KUMAR*, ARMELLE WAUTIER**

** Alcatel, Corporate Reasearch Center, Radio Dpt., 5 rue Noël Pons, 92734 Nanterre, France*
*** Ecole Supérieure d'Electricité, Dpt. Radio-Electricité, Plateau du Moulon, 91192 Gif sur Yvette, France*

Abstract: Several multi-dimensional trade-offs between coverage area, capacity, quality of service, required bandwidth and cost need to be considered for the deployment of cellular networks. Typically, large cells (radius of several kilometers) guarantee continuous coverage in low traffic service areas, while small cells (radius less than 1 kilometer) are deployed to achieve higher capacity. Due to the tremendous success of cellular systems network planning to cater for the traffic capacity requirements of "hot spots" becomes a critical issue. Techniques such as deployment of small cells (micro-cells) and efficient management of radio resources are used to manage high traffic density with limited available spectrum bandwidth. In TDMA cellular systems such as GSM (900 or 1800 MHz), PCS 1900 or D-AMPS, reduction in cell size means a more frequent spatial reuse of frequencies and hence a higher spectral efficiency. However, the increasing difficulty of ensuring good quality handovers with decreasing cell sizes imposes an asymptotic limit for this method of performance enhancement. This chapter, first describes the "conventional methods" for capacity enhancement of TDMA based cellular systems and then develops the principle of hierarchical networks useful for very high density networks. It corresponds to a network organization where at least two different cell types (e.g. macro-cells and micro-cells) operate in an overlapping coverage and employing special means of interlayer resource management (directed retry). Finally, the idea of "distributed coverage" in the micro-cell layer is introduced. It is demonstrated that the communication quality is improved, offered traffic is increased and the accuracy of mobile speed estimation is also enhanced, further improving the spectrum efficiency in the service area.

1. PRINCIPLES OF RADIO CELLULAR NETWORK DESIGN

The design of a cellular network is based on analysis of trade-offs between several parameters of the base station sub-system (BSS). The major objective is to serve a maximum number of mobile subscribers with acceptable quality. The following paragraph presents the quality metrics and other parameters involved in this process.

1.1 Quality of service and grade of service

The quality of service (QoS) of a cellular network, perceived by the users, depends upon call quality and network availability. Moreover, call continuity and quality of handovers are other important considerations.

In-call speech quality is usually measured by the mean opinion score (MOS) value that ranges between 0 (very bad quality) and 5 ("hi-fi" quality). The MOS is a consistent and worldwide accepted subjective criterion but it is difficult to assess or predict in an operational network. More manageable (i.e. objective) performance criteria for digital information transmissions (corresponding to voice or data) are the bit error rate (BER) or frame error rate (FER). For an acceptable operation, BER and FER have to be maintained below some predetermined threshold values. The actual BER and FER depend on the transmission parameters (source coding, channel coding, interleaving and modulation) and on the propagation environment. The bit error rate performance threshold can be translated into a minimum required signal to noise ratio (SNR) depending on the air-interface parameters and on the power-delay profile of the channel. This SNR threshold is around 9 dB for GSM.

Network availability consists of two parts, good quality radio coverage, and availability of enough radio resource (communication channels) on the base station. Generally speaking, sufficient radio signal strength needs to be provided over 90% to 95% of the cell coverage area so that the received BER / FER can be maintained below quality threshold. Margin to compensate for lognormal shadowing (slow fading) has to be duly considered. Further, cell by cell calculation of link budget, to ensure balanced link (uplink and downlink) is performed. Finally, the selected frequency reuse pattern for network deployment has to be such that only a controlled amount of co-channel interference is generated. This latter depends upon the path-loss model, cell geometry, number of active mobiles and their location.

As far as the resource availability is concerned, the quality can be expressed by the number of calls that are rejected or blocked at connection set-up. Teletraffic models can be used to calculate the call blocking probability. In a frequently used traffic model for voice services, call arrivals are modeled according to a Poisson random process with a call rate arrival denoted by λ (calls per second). For cellular networks, λ is relative to a given area. Call duration is assumed to be exponentially distributed with an average duration of T_m seconds. The offered traffic ρ expressed in Erlang is simply the product $\lambda\ T_m$. Blocking probability p_b is the probability that all the servers (channels) are loaded. Loss probability depends on the offered traffic, on the number of channels, and on the resource management policy. Let us consider that a call is lost only when all the radio resources assigned to the cell, where the mobile attempts to initiate its call, are fully loaded. In that case, the loss probability p_l is the probability that all the channels are fully loaded while a new call arrives; and loss and blocking probabilities are equivalent. The Erlang B formula (cf. equation [1]) gives the blocking rate p_b as a function of the offered traffic ρ and of the number of radio resource for traffic per cell M.

$$p_b = \frac{\dfrac{\rho^M}{M!}}{\sum_{k=0}^{M} \dfrac{\rho^k}{k!}} \qquad [1]$$

Usually, a blocking probability target of 2 % is considered when designing cellular outdoor systems.

In a radio mobile network, a call may also be dropped during a handover procedure (when, for instance, no channel is available in the target cell or when the SNR goes below the SNR value tolerated by the receiver). This causes a forced call termination, which is much less tolerable than a blocked call. The dropped call probability is very sensitive to mobile speed versus cell size and to radio resource management strategies (handover parameters and associated algorithms).

The loss probability p_l and dropped call probability p_d are usually grouped into a single performance criterion called the GoS (Grade of Service). GoS is an objective criterion reflecting both the network availability and the efficiency of radio resource management. It is defined by:

$$GoS = p_l + 10 p_d \qquad [2]$$

1.2 Design and Dimensioning of Cellular Networks

Fundamental parameters for network design / dimensioning are:

- total coverage area and terrain topology,
- traffic density and its variance
- required probability of good coverage and the associated SNR and signal to interference ratio,
- GoS (including the effect of blocked and dropped calls).
- The design process (consisting of some iterations) is aimed at providing acceptable quality of service to maximum number of users at minimum expense in radio spectrum and in number of cell sites. Models of user activity (traffic and mobility patterns) and those for signal and interference propagation are duly considered in the process. The final outcome is given in terms of:
- cellular structure,
- number of cells / sites to cover the service area,
- radius of each cell,
- number of channel elements per cell,
- frequency reuse pattern for traffic and beacon frequencies,
- strategy for resource allocation and for handover in the BSS.

In the early phase of a cellular network deployment, macro-cells are used. A macro-cell may have large coverage range (up to few tens of kilometers). In practice, the coverage area is linked to transmitted power and to the antenna height. Low traffic areas are covered with large macro-cells (radius of several kilometers and high antennas) while dense traffic areas are covered with smaller macro-cells (radius of several hundred metres).

In TDMA cellular systems, fixed channel allocation (FCA) is generally used. A predetermined number of radio frequency carriers are assigned to each cell. The number of channel elements depends on the assigned number of carriers and on the number of time-slots per carrier. Table 1 shows an example of calculation of the offered traffic (in Erlang, for a call blocking of 2 %) versus number of assigned carriers in the cells of a GSM network. This calculation does not take into account mobility (handovers are not considered) and assumes that all the unused traffic channels are always available in the resource allocation procedure. However, in dense traffic areas, where small cells are deployed, there is an increase in the average number of handovers per call. The probability of dropped calls during handover (due to unavailability of resource in the target cell) tends to increase and it needs to be addressed when calculating the GoS.

Number of frequency per cell	1	2	3	4	5	6	7	8
Time slot for beacon information	1							
Time slots for dedicated signalling	1		1		1		1	
Time slots for traffic	6	8	7	8	7	8	7	8
Cumulated number of traffic channels	6	14	21	29	36	44	51	59
Offered traffic for 2% blocking probability (ρ Erlang / cell)	2.28	8.2	14	21	27.3	34.7	41.2	48.7
Efficiency: ρ/M	0.38	0.58	0.66	0.72	0.76	0.79	0.81	0.82

Table 1. Offered traffic vs. number of carriers per cell with GSM air-interface

An important step in cellular network design is the selection of a frequency reuse pattern. If the traffic density is uniform for the whole service area, cell size can be identical and the number of carriers per cell as well. In this scenario, the frequency plan may be periodic with a reuse factor of N: the available frequencies are allocated in N cells forming a cluster, and the same cluster is repeated in the service area. The choice of N is related to the acceptable signal to interference ratio. The total number of available carriers (divided by the reuse factor) limits the maximum number of carriers per cell. The reuse factor N is usually large in a first phase and it needs to be reduced when the traffic per cell increases. For existing FDMA / TDMA networks, typical values for N are 21, 18, 12, and 9.

2. CONVENTIONAL WAYS TO ENHANCE TRAFFIC CAPACITY

2.1 Solutions for Macro-cells

In a traditional macro-cellular network, the capacity enhancement is obtained by increasing the number of carrier frequencies per base transceiver station (BTS). This is the most straightforward method, but the achievable capacity enhancement is clearly limited by the total allocated spectrum and by the frequency reuse pattern N. Nevertheless, this capacity increase does not affect the quality of service since both the coverage and the frequency reuse factor remain unchanged (if the additional frequencies are taken from the same frequency band). However, it may happen that the additional spectrum comes from a different band (for instance a GSM 900 MHz operator gets a licence for some frequencies in the 1800 MHz band). In this case, the network upgrade requires additional inter frequency band handover mechanisms and a different frequency planning.

Then, further capacity gain can be obtained by implementing more compact frequency reuse patterns. The subsequent degradation in transmission quality due to the increase in interference level can be mitigated by implementing interference control techniques such as power control (PC), voice activity detection and discontinuous transmission (VAD/DTx) or slow frequency hopping (SFH) [Verhultz, 90], [Nielson, 98]. With these techniques a capacity gain between 20 to 30 % is achievable. A more significant capacity gain can be obtained by using spatial division multiple access (SDMA) techniques. The cluster size can be reduced by a factor 3 using adaptive beam forming and interference cancellation mechanisms with antenna arrays [Kuchar, 99]. However, these techniques cannot be applied to the frequencies carrying beacon or common control signals. Consequently, different frequency reuse patterns are used inside the network: one for beacon frequencies and another one, more compact, for traffic channels.

The characteristics of the voice codec used for transmission of the signal on the air-interface are also paramount importance to determine the network capacity (cf. §1.1). Indeed, the voice codec rate as well as the associated channel encoder determine the amount of radio resource necessary for one communication but also the required SNR for obtaining suitable BER and FER figures. In turn, the SNR determines the frequency reuse factor.

In the case of full rate speech codecs (FR), a communication occupies one time slot per TDMA frame on one frequency. In the case of half rate speech codecs (HR), two communications may be time-multiplexed on the same radio resource. Consequently, the usage of half rate voice codecs in the BSS may double the number of traffic channels per carrier and therefore increase the capacity by more than a factor 2 (due to trunking efficiency). However, the requirement of higher SNR (for similar voice quality as FR) has hindered the deployment of HR and these have been advantageously replaced by adaptive multiple rate voice codecs (AMR). AMR codecs offer dynamic adaptation between HR / FR modes as well as dynamic adaptation of source / channel coding inside the codec modes. This adaptation is related to receive SNR. The "adaptation" leads to an extension in range of operational SNR which can be leveraged for a more compact frequency reuse and hence an increase in traffic capacity [Corbun, 98].

Further capacity enhancement can also be obtained by reducing the cell size. This can be realized by cell splitting or by inserting new transmission sites in the network. Cell splitting consists in replacing omni-directional cell sites by sectored cells and by adjusting the corresponding appropriate reuse pattern. Adding new sites is another solution to reduce cell area. In this case, the maximum coverage distance must be reduced, and antenna height and/or transmit power should be decreased.

2.2 Deployment of Micro-cells

In areas with very high traffic demand, (for instance downtown city centres, shopping areas, airport, etc...) very small cells have to be deployed. The radiated energy is kept confined to small areas by installing the (micro-) cell site antenna below the rooftops of surrounding buildings. The propagation environment thus obtained is quite different form the one for macro-cells. Indeed, in micro-cellular environment, the propagation is guided along the streets [Xia, 94], [Andersen, 95]. The signal power decreases slowly with the distance between the mobile and the base station d when the mobile is in line of sight (LOS) (about $1/d^{2.6}$ before the "breakpoint" and $1/d^4$ after the "breakpoint"). There is an abrupt power loss (about 20 dB over a few metres) when the mobile turns at a street corner (i.e. the mobile goes in a non line of sight condition, NLOS). In order to secure the handover operation, the received signal strength has to be maintained above a certain threshold even in NLOS condition. This leads to a substantial coverage overlap between the adjacent cells for the LOS paths (cf. figure 1). For an MS in the overlap region, there is a small power difference between beacon signals coming from two adjacent cells in LOS. This makes handover tuning much more difficult than in a macro-cellular system.

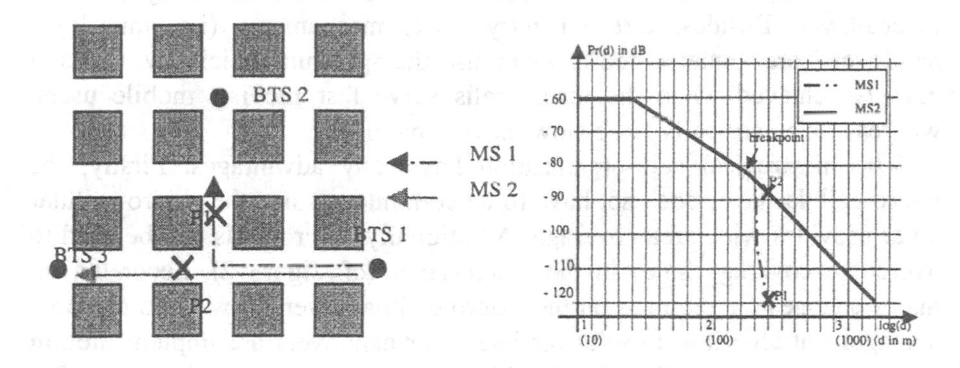

Figure 1: Path-loss variations for mobiles in NLOS (MS1) and LOS (MS2) from BTS1

Besides, co-channel interference coming from LOS cells is increased too. These problems can induce an increased probability of "too late handover decisions" or even "wrong target cell selection". The situation is worsened by the high value of standard deviation of the lognormal shadowing.

Moreover, since the average number of handovers per call also increases, it very difficult to maintain the dropped call rate below an acceptable value. So, in a micro-cellular environment, traffic enhancement is clearly limited by micro-cellular specific propagation condition and by the average number of handovers (i.e. related mobile speed).

In fact, some of the challenges in ultra dense traffic areas (interference and handover management) can be met by using a hierarchical multi-layer cellular network organization (cf. figure 2). The micro-cells are deployed as an underlay of the macro-cells.

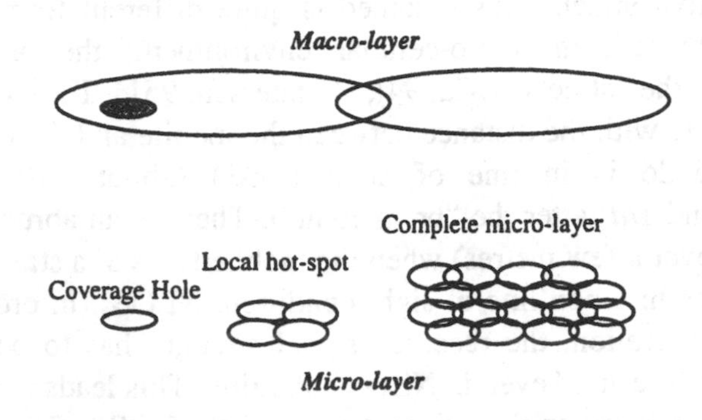

Figure 2: Hierarchical multi-layer network and deployment scenarios

In the hierarchical structure, resource allocation mechanisms in the BSS must determine whether to connect new calls to the micro-layer or the macro-layer. Besides, directed retry (DR) mechanisms (i.e. inter-layer handovers) are implemented to maximise the spectrum efficiency. Optimal gain is achieved when the macro-cells serve fast moving mobile users, whereas the micro-cells serve slow moving mobiles.

The hierarchical cell organization has many advantages. Firstly, the micro-cellular layer does not have to be continuous since the macro-cellular layer provides wide area coverage. Additionally, micro-cells can be used to overcome coverage holes in the macro-layer (cf., figure 2). Secondly, no major changes are required on the macro-cellular layer. New radio resource management algorithms useful for interlayer handovers are implemented in the base station controller (BSC). Thirdly, geographical variations in traffic densities can be handled with good overall spectrum efficiency in the network.

3. DISTRIBUTE COVERAGE: A NOVEL CONCEPT FOR VERY HIGH DENSITY AREAS

The capacity limitations of conventional micro-cellular can be resolved by adopting an original organization of the BSS based on a distributed coverage.

3.1 Network Architecture and Main Features

Here, the conventional BTSs are replaced by a set of synchronous radiating elements, the relays, that are connected to a local control equipment, the base transceiver station concentrator (BTSC). The cell is therefore defined by the coverage of the set of relays connected to one BTSC (cf. figure 3). The relays are deployed so that there is no coverage hole in the cell. With this solution, the number of sites is increased (each site having a reduced radio coverage) while the size of the cells is kept to an acceptable value. From a functional point of view, the BTSC and associated relays are seen as a conventional BTS from the BSC. With multi-element coverage, the cell dimension are large enough so as to permit an acceptable operation of inter cell handovers. No additional expenses related to beacon carriers needs to be incurred.

The multi-site illumination inside the cell improves the coverage quality due to "simulcast" (transmit diversity) in the downlink and to macro-diversity in the uplink (receiver diversity). Therefore, lower transmit power relay can be used, which reduces the coverage overlap between cells. Examples to demonstrate the capacity gain of such an architecture are available [Ariyavisitakul, 96].

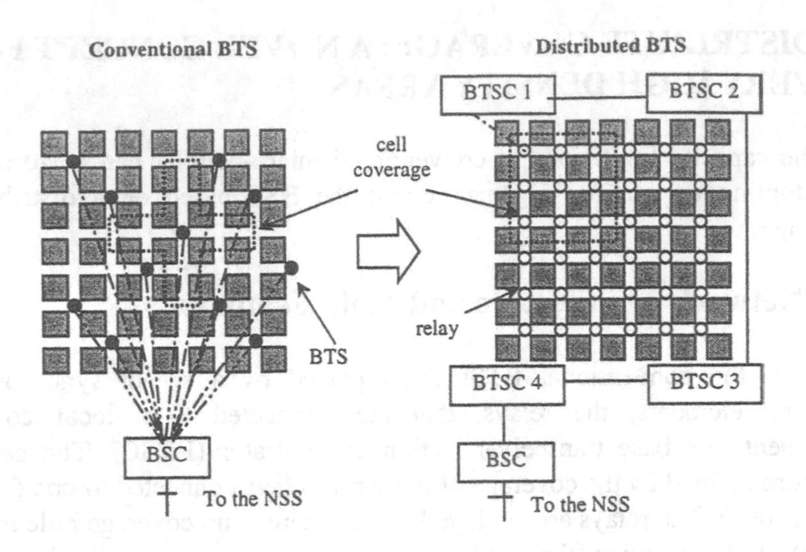

Figure 3: Conventional BTS vs. distributed BTS

With this type of BSS architecture, additional significant gain can be obtained by radiating signal energy on a selected set of relays in the cell. A procedure to determine the "set of best server relays" needs to be implemented in the BTSC. A regular updating of this "set of best server relays" is essential for moving subscribers. Such updating procedures can be based on uplink and/or downlink measurements. Convergence time of typically 100 ms is achievable [Charrière, 97]. Generally speaking, a moving subscriber has to be offered the possibility to use the same physical channel (i.e. a specific couple time slot and frequency or hopping sequence) throughout the communication. The activation / de-activation of "same physical channel" on the evolving set of best server relays is realized according to a procedure termed ACT (Automatic Channel Transfer). The ACT differs from conventional handover mechanisms. Indeed, ACT does not require any exchange of signalling messages between the network and the mobile terminal and does not need connection release and re-establishment on a new physical channel. Therefore, ACT has a positive impact on the perceived communication quality (no interruption of traffic flow) and also on the network performance (the probability of forced call termination is reduced). Several different strategies for selection between conventional and handover and ACT are available [Wautier, 98].

3.2 Equipment Architecture

In a distributed BSS, the BTSC and its associated relays perform the same functions as a BTS. This includes:

- broadcast of beacon and common control messages,
- management of new call requests,
- transmission/reception of traffic and signalling messages on the air-interface,
- measurement reporting to the BSC for mobility management.

In addition, the BTSC must perform selection of best server relays and ACT.

As shown in figure 4, simplified relays include the basic transmit and/or receive functions for broadcast carrier (B-TRx), traffic carrier (P-TRx). The optional scan receiver (Scan-Rx) function in the relay is helpful in speeding up the procedure for best server selection.

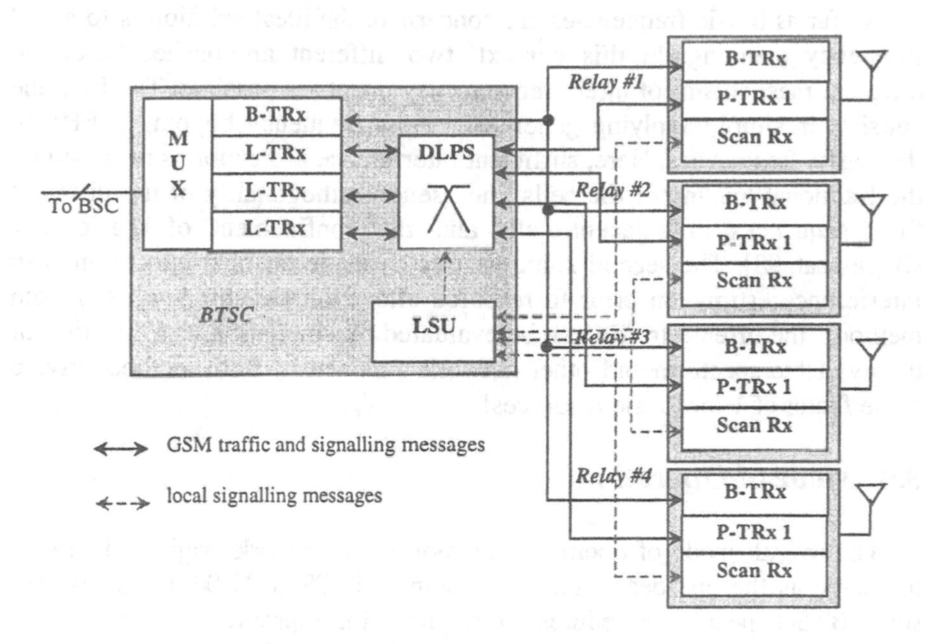

Figure 4: Functional architecture of a distributed BTS

The BTSC implements:

- transceiver function for the beacon frequency (B-TRx),
- a set of transceivers functions for traffic channels (L-TRx),
- a switch, termed DLPS, to enable dynamic interconnection between
  L-TRx and P-TRX,

- a local selection unit (LSU) that performs local resource allocation and best server relays selection, and also drives the DLPS.

3.3 Frequency planning

Frequency planning for beacon and traffic is handled rather independently.

Beacon frequencies must transmit at a constant power level on every time slot, including the ones that do not convey any beacon information. Hence, no energy confining technique can be applied to such frequencies. The set of beacon frequencies must be allocated using FCA in a conventional way. Dedicated control channels should also be assigned on the beacon frequency to ease the best server relay selection at call establishment. Some spare time slots might also be reserved on these carriers for securing incoming conventional handovers (cf. §3.4).

As far as traffic frequencies are concerned, the ideal solution is to avoid frequency planning. In this context, two different approaches based on different mechanisms of interference management are possible. The first one consists in simply applying generalised slow frequency hopping (SFH) on the traffic frequencies. Here, sufficient interference protection is provided by the fractional loading of the cells, the pseudo-orthogonality of the different SFH sequences in adjacent cells and the confinement of the energy [Bégasssat, 98]. The second solution uses dynamic channel allocation with interference estimation prior to resource allocation [Kazmi, 99a]. For both methods, the offered traffic can be evaluated by simulation as a function of the available spectrum and other network parameters. Both methods give a reuse factor of 1 for traffic resources!

3.4 Mode of Operation

The overall mode of operation for mobile or network originated calls is the same as the one described in the standard [TS GSM 04.01]. However, some BTSC specific procedures are required for appropriate call operation. Here, only the solution using generalised frequency hopping is described. For further explanation on the other possible solution, the reader is referred to [Kazmi, 99a].

Call establishment and resource allocation

As in a conventional GSM network, the different phases of call establishment and resource allocation procedure involve the exchange of signalling messages on the beacon frequency.

For a mobile originated call, a request is sent to the network on the "random access channel". This burst is received and demodulated by several relays but with different power and quality levels. An access grant message is sent back to the mobile terminal using the "common control channel". Then, a dedicated control channel is allocated by the network. Measurements for the selection of the best server relays can be performed during the exchange on the "dedicated control channel". Typically, at least two relays are "kept active" for every on-going call (cf. figure 5). A good uplink and downlink quality and high probability of keeping every mobile in LOS with respect to at least one relay are thus maintained.

The activated radio resource on best server relays is defined on the basis of time slot number (in the TDMA frame), hopping sequence number (HSN) for generalised SFH and its offset index. Careful selection of HSN and the offset index is essential to maintain sufficient orthogonality between on-going calls (using a given time slot). The latter can be optimized through HSN allocation compatible with good operation of ACT for mobile subscribers. For more details, the reader is referred to [Wautier 98].

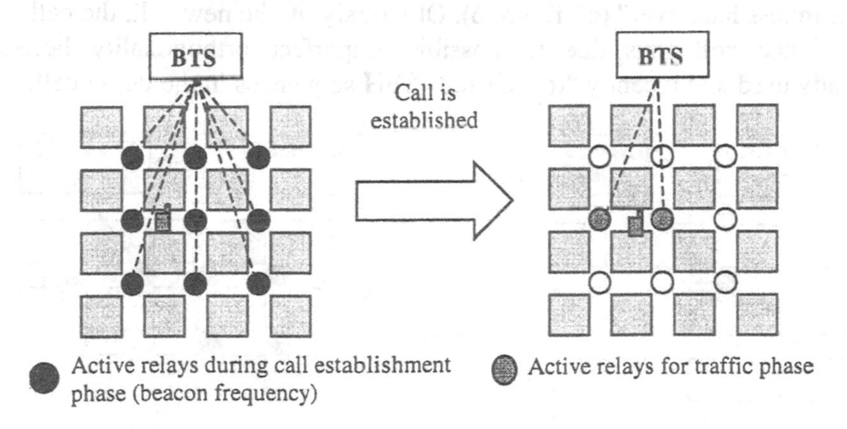

Figure 5: Call establishment procedure

Mobility Management during Communications

One should distinguish between two families of mobility management procedures in this BSS architecture:

- automatic channel transfer (ACT),

- conventional handover.

ACT mainly deals with intra-cell mobility management and consists in the dynamic selection of the best server relays for the mobile. During this procedure, same physical channel is maintained. ACT procedure are handled by the relays / BTSC independently of any involvement of the mobile. The "scan receiver" can be used to perform uplink signal level and quality measurements useful to determine the "best server relays" in the immediate vicinity of a mobile subscriber. Exchange of signalling information between BTSC and the relays as well as an appropriate setting of the DLPS (cf. figure 4) are essential for functioning of ACT procedures. An evaluation of this local signalling can be found in [Kazmi, 99b].

ACT may also be performed during inter-cell mobility management procedure. Time synchronisation of all the TRx functions if relays in different cells is necessary. The BTSC correlates the uplink measurements with the standard measurement reports that are sent by the mobile to the network (i.e. received power levels from neighboring cells). The current serving cell can initiate a "best server relay selection procedure" in the target cell for the considered communication. Once this operation is performed and if corresponding physical channel is available in the target cell (i.e. the same time slot is free on at least one transceiver of the target best server relays), the SFH sequence parameters can be transferred to the target BTSC and the communication can continue without any change of physical channel: this is a "seamless handover" (cf. figure 6). Obviously, in the new cell, the call may experience collisions due to possible imperfect orthogonality between already used and recently "transferred" SFH sequences in the target cell.

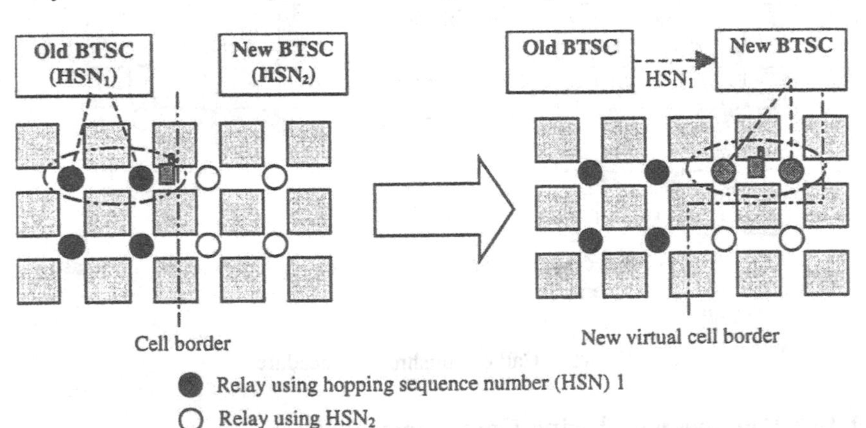

Figure 6: "Seamless handover" and virtual cell border

Conventional handovers may occur during intra-cell mobility management when the same physical channel is not available on the new set of best server relays. This is executed via exchange of some local signalling messages between the BTSC and the terminal. Allocation of a new time slot, SFH sequence and offset index completes the procedure.

Besides, conventional handovers might be required when a user moves from one cell to another and if inter-cell ACT is not feasible. This can be handled by a one-phase or a two-phase handover. In one phase handover, the network takes benefit from geographical knowledge of network topology. The related BTSC immediately activates the relays of the target cell that are closest to the ones that were previously active in the old serving cell. In the second approach, a more secure mechanism requiring 2 consecutive handovers is implemented. The call is first handed over an available traffic channel on the beacon frequency. Therefore the call is active on all the relays on the new serving cell for a short duration. Selection of best server relays and subsequent radio resource allocation on a traffic frequency can be performed.

3.5 Deploying distributed BTS in Hierarchical Networks

The distributed BTSs can be deployed as a mono-layer network covering a service area with very high traffic demand. Such a solution can also be advantageously deployed as the lower layer of a hierarchical network (cf. figure 7). In fact, the distributed coverage is also very helpful in improving the efficiency of interlayer mechanisms. This is particularly true for the DR procedure.

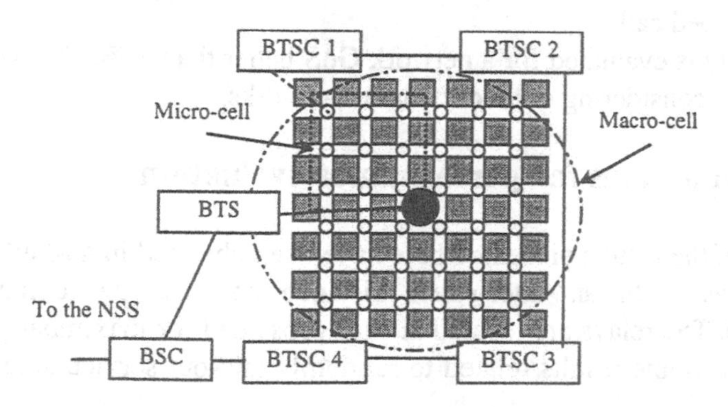

Figure 7: Hierarchical network with lower layer based on distributed coverage

In the hierarchical scenario, new calls are preferably initiated on the micro-layer. Simple, accurate and fast measurement of displacement rate of mobile users is available as a by-product of the ACT procedure. Consequently, "fast moving" users can be re-directed to the upper-layer and the traffic capacity / quality of service can be optimized.

Two possible criteria (based on ACT) for redirecting calls between network layers are briefly described below. In the first case (criterion C1), a call is handed off to the macro-layer if the number of ACTs that have been performed for this call exceeds a predefined threshold, N_{ACT}. The second criterion, C2, is related to the availability of the best server relays: if there is no resource available to provide coverage from more than 1 relay for a call, it is handed off to the macro-layer. The corresponding merits and demerits of these criteria and the corresponding performance figures are presented in the next section.

4. PERFORMANCE OF A DISTRIBUTED BTS BASED BSS

The performance of the proposed network organization has been analysed through simulations. For this purpose, a C++ event driven tool has been developed. This simulator includes:

- environment parameters (relay locations, street and terrain topology, ...),
- traffic models,
- mobility models,
- a set of radio resource management algorithms (allocation, mobility),
- computational sub-routines to calculate the probability of blocked and dropped call.

Capacity is evaluated for a network GoS better than 2 %. This is a typical value when considering outdoor cellular networks.

4.1 Assumptions for performance evaluation

Most of the results presented here have been obtained in a Manhattan like service area with an inter-street distance equal to D (default value, D=100 m). The relays are located at each crossroad for maximising the LOS probability. Some results related to randomly laid-out service areas are also available.

In the streets, a conventional micro-cellular propagation model is considered, that is a 2-slope path loss model for LOS and a secondary source model plus corner effect for NLOS [Xia, 94]. An interference free scenario is considered. The availability of a free resource is a function of the distance between the mobile and the relay and of the propagation parameters. This situation is realized through fractional loading. Further details on this issue are provided in [Bégassat, 98] where interference level is evaluated as a function of fractional loading and other system parameters.

Only voice applications are considered. The commonly used model based on a Poisson distribution of the inter-arrival instants and on a negative exponential distribution of the call duration is simulated.

The simulated population consists in a mix of fast and slow users. Here, the mobile speed is normalised with respect to the distance between two adjacent relays. A slow user is defined as a user that stays in the same zone (distance between two adjacent relays) throughout the call while a fast user might cross several zones during the call. In the following, the speed of fast mobile users is denoted by h i.e. the average number of zones that are crossed during a call. A quasi random mobility pattern for the fast mobiles is assumed. In fact, at crossroads, there is an equal probability (that is 1/3) for a mobile to go straight on, turn left or turn right.

The performance figures presented in this paper are related to a time slot allocation strategy whereby the time slot available on the maximum number of relays of the BTSC area is allocated to the incoming calls. Detailed analysis of several other strategies is available in [Wautier, 98].

4.2 Performances of Mono-layer Deployment

A range of parameters, listed below, have been considered in performance evaluation:

- number of relays per BTSC,
- number of P-TRx per relay, $N_{P\text{-}TRx}$,
- number of available RF frequencies,
- number of L-TRx per BTSC,
- number of available hopping sequences N_{Seq},
- updating period of list of "best server relays".

Figure 8 is an example of traffic variation of offered traffic per relay as a function of the number of available slow frequency hopping sequence (N_{Seq}) for different relay configurations.

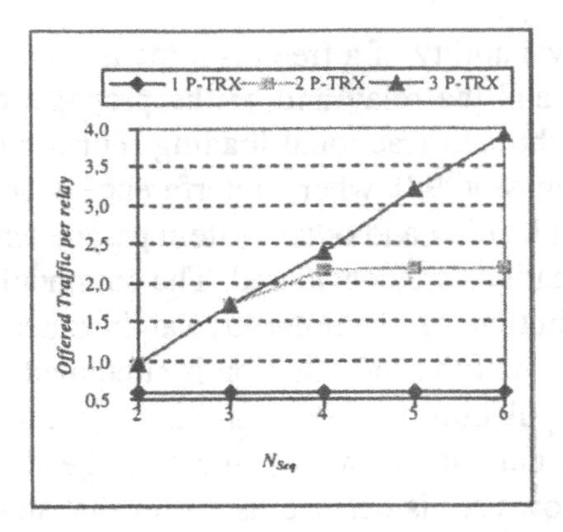

Figure 8: Offered traffic per relay (in Erlang) vs. $N_{P\text{-}TRx}$ and N_{Seq}

The traffic mix between fixed and moving users as well as the speed of moving users are also important parameters affecting the traffic capacity. This is illustrated in figure 9.

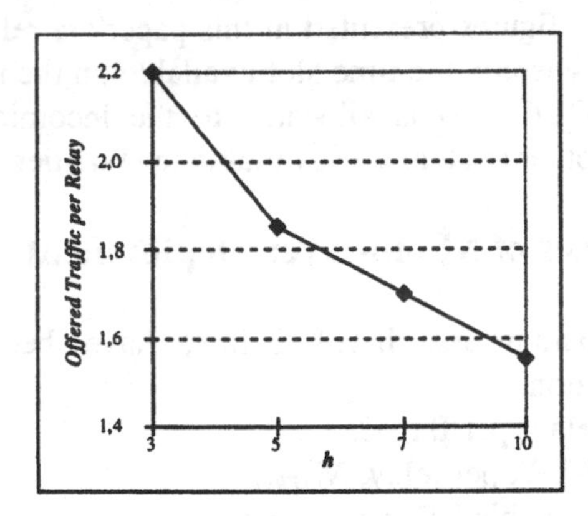

Figure 9: Offered traffic (in Erlang) vs. normalised speed of moving users

This shows that the utilization of mono-layer distributed BTS based BSS should be limited to pedestrian areas where user mobility is low. In such a case, more than two fold increase in spectrum efficiency (as compared to conventional mono-layer micro-cellular network deployment) is achievable.

4.3 Performance in a Multi-layer Deployment

A two layer network consisting of nine micro-cells with 9 relays per micro-cell and a single overlay macro-cell has been simulated (cf. figure 10). To eliminate any "border effect" induced by the mobile subscribers leaving the two layer coverage area, a continuous coverage by the macro-cellular layer is assumed.

Each relay carries on B-TRx and one P-TRx capable of handling 8 simultaneous traffic channels (corresponding to an 8 time slots TDMA system). Each BTSC controlling 9 relays of a micro-cell is able to manage N_{Seq} L-TRx (corresponding to 8xN_{Seq} simultaneous calls).

Performance curves showing the influence of certain significant system parameters are presented in figure 11 and 12. Following parameters have been considered:

- Criteria C1 and C2 for DR,
- N_{ACT}, the maximum permissible number of ACT before DR according to C1 is initiated,
- N_{Macro} the number of carriers available for the macro-cellular layer (each carrier offers the capability to handle 8 simultaneous calls).

It can be observed that:

- A two layer approach (with distributed coverage in the lower layer) offers more than three times increase in traffic throughput as compared to a single layered network (cf. figure 11 and 12),
- A value N_{ACT} equal to 2 provides optimum throughput in case of DR based on criterion C1. This is related to the timely execution of DR i.e. neither too early to over load the macro-layer and nor too late such that an ongoing call is prematurely curtailed.
- Traffic throughput performance for DR based on criterion C1 degrades with increasing population of high speed mobiles (cf. figure 12).
- Application of DR based on criterion C2 ensures good traffic throughput for all types of subscriber mobility profile.

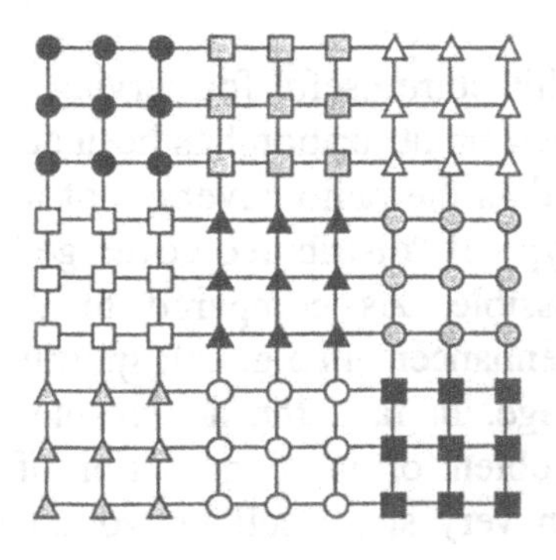

Figure 10: Simulated service area: 1 macro-cell + 9 micro-cells

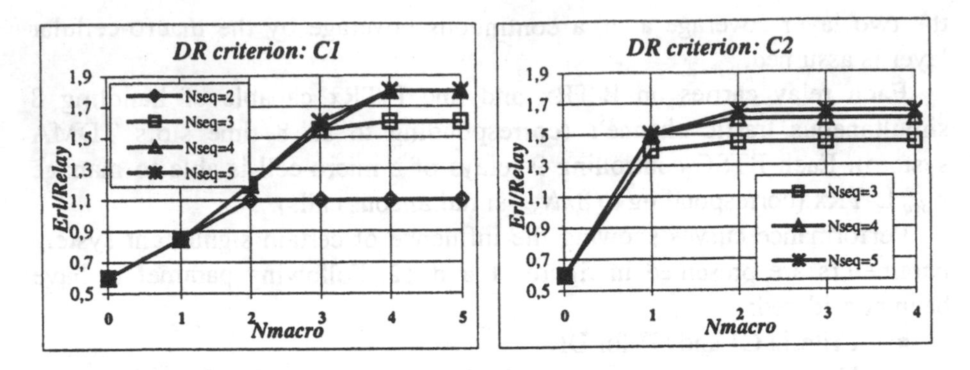

Figure 11: Offered traffic per relay on the micro-layer vs. N_{seq} and N_{Macro} for h=3 (low speed)

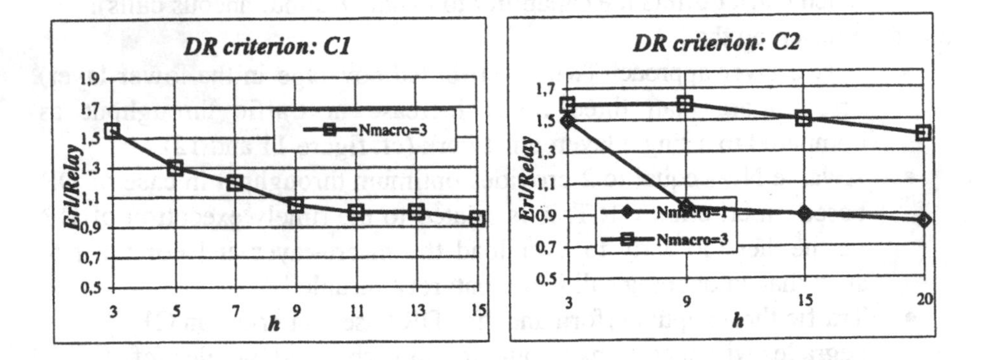

Figure 12: Traffic capacity vs. normalised mobile speed (N_{Seq}=4)

5. CONCLUSION

A novel cellular architecture useful for serving very high traffic density areas with sufficient spectrum utilization has been described. In the proposed scheme, a cell is defined as the radio coverage of a set of relays controlled by a BTSC. With this type of "distributed coverage", fine adjustment of the cell dimensions is possible. As compared to the conventionally used techniques for capacity enhancement e.g. cell splitting, the proposed solution offers a double advantage. In fact, the achievable increase in capacity is much higher and the problem of any degradation of GoS (due to increased number of handovers in very small cell networks) is completely avoided. Moreover, the deployment of distributed coverage in the "lower layer" of a hierarchical network is found to be helpful in improving the effectiveness of directed retry and radio resource allocation according to the mobility profile of the subscribers. Further advantages, as compared to cell splitting, are the savings in beacon frequency assignment and more flexibility in frequency assignment. In fact, all the relays in a cell use the same beacon frequency.

The advantages are somewhat counter-balanced by the increase number of "points of transmission" in a cell. However, a relay needs to radiate only o few mw of power. Hardware for traffic and beacon carriers and the antenna can easily be integrated in the existing urban equipment (lamp post, etc, ...).

Performance results for a GSM based outdoor network for speech services have been presented. These can be easily extended to other TDMA systems. The applicability of "distributed coverage" to third generation systems has also been studied [Mihailescu, 99]. The techniques presented in this paper are also applicable to ensure continuous coverage in different environment (indoor to outdoor) as well as for throughput enhancement in applications with variable bandwidth allocation [Brouet, 99], [Kazmi, 00].

REFERENCES

[Andersen, 95], Andersen J.B., Rappaport T., Yoshida S., "Propagation Measurements and Models for Wireless Communication Channels", *IEEE Communication Magazine*, January 1995.

[Ariyavisitakul 96] Ariyavisitakul S., Darcie T.E., Greenstein L.J., Philips M.R., Shankaranarayanan N.K., "Performance of Simulcast Wireless Technique for Personal Communication Systems", *IEEE Journal on Selected Areas on Communications*, vol. 14, pp. 632-643, May 1996.

[Bégassat, 98] Bégassat Y., Kumar V., "Interference Analysis in an Original TDMA-based High Density Cellular Radio Network", *Proceedings of VTC'98*, Ottawa, May 1998.

[Brouet, 99] Brouet J., Nousbaum F., "Performance of a Self-organising GSM based System with Distributed Coverage for High Density Indoor Applications", *Proceedings of VTC 99*, Amsterdam, Sept. 1999.

[Charrière 97] Charrière P., Brouet J., Kumar V., "Optimum Channel Selection Strategies for Mobility Management in High Traffic TDMA-based Networks with Distributed Coverage", *Proceedings of ICPWC'97*, Bombay, Dec. 1997.

[Corbun, 98]. Corbun O., Almgren M., Svanbro K., "Capacity and Speech Quality Aspects Using Adaptive Multi-Rate (AMR)", *Proceedings of IEEE PIMRC'98*, Boston, Sept. 1998.

[Dreissner, 98] Dreissner J., Barreto A.N., Barth U., Feittweis G., "Interference Analysis of a Total Frequency Hopping GSM Cordless Telephony System", *Proceedings of IEEE PIMRC'98*, Boston, Sept. 98.

[Kazmi, 99a] Kazmi M., Godlewski P., Brouet J., Kumar V., "Performance of a Novel Base Station Sub-system in a High Density Traffic Environment", *Proceedings. ICPWC'99*, Jaïpur, Feb. 1999.

[Kazmi, 99b] Kazmi M., Brouet J., Godlewski P., Kumar V., "Handover Protocols and Signalling Performance of a GSM based Network for Distributed Coverage", *Proceedings of VTC'99-Fall*, Amsterdam, Sept. 1999.

[Kazmi, 00] Kazmi M., Brouet J., Godlewski P., Kumar V., "Radio Resource Management in a Distributed Coverage Mobile Multimedia Network", *Submitted to PIMRC 2000, Sept. 2000, London.*

[Kuchar, 99] Kuchar A., Taferner M., Bonek E., Tangemann M., Hoeck C., "A Run-Time Optimized Adaptive Antenna Array Processor for GSM", *Proceedings of EPMCC'99*, Paris, March 1999.

[Mihailescu, 99] Mihailescu C., Lagrange X., Godlewski P. "Locally Centralised Dynamic Resource Allocation Algorithjm for the UMTS in Manhattan Environment", *Proceedings of PIMRC'98*, Boston, Sept. 1998.

[Nielsen, 98] Nielsen T.T., Wigard J., Skjaerris S., Jensen C.O., Elling J., "Enhancing Network Quality Using Base-band Frequency Hopping Downlink Power Control and DTx in a Live GSM Network", *Proceedings of IEEE PIMRC'98*, Boston, Sept. 1998.

[TS GSM 04.01] "MS-BSS Interface –General Aspects and Principles", ETSI.

[Verhulst, 90] Verhulst D. "High Performance Cellular Planning with Frequency Hopping", *Proceedings of the Fourth Nordic Seminar on Digital Land Mobile Radio Communications*, Oslo, June 1990.

[Xia, 94] Xia H.H. et al, "Micro-cellular Propagation Characteristics for Personal Communications in Urban and Suburban Environments", *IEEE Transaction On Vehicular Technology.*, vol. 43, n°3, August 1994.

[Wautier, 98] Wautier A., Antoine J., Brouet J., Kumar V., "Performance of a Distributed Coverage SFH TDMA System with Mobility Management in a High Density Traffic Network", *Proc. PIMRC'98*, Boston, Sept. 1998.

Chapter 8

TRAFFIC ANALYSIS OF PARTIALLY OVERLAID AMPS/ANSI-136 SYSTEMS*

R.RAMÉSH AND KUMAR BALACHANDRAN

Ericsson Research, Research Triangle Park, NC

Abstract: The problem of calculating the traffic allowable for a certain grade of service in a cellular network employing both AMPS and ANSI-136 channels is considered. The dual-mode capability of the ANSI-136 users enables the system to assign them to AMPS channels if ANSI-136 channels are blocked; the two pools of users cannot be treated independently. An analytical method for the calculation of the traffic is derived and the actual capacity improvements obtained by a partial deployment of ANSI-136 are shown. The chapter derives a strategy to maximize the number of ANSI-136 users supported for a given number of AMPS users. The case of reconfigurable transceivers at the base station is also considered and the allowable traffic derived. It is seen that a significant increase in traffic can be achieved by this option, albeit at the price of increased system complexity.

* Parts of this work were presented by the authors at PIMRC'98.

1. INTRODUCTION

The ANSI-136 system was conceived as a natural evolution of AMPS for higher capacity and provides cellular operators with an option of significant backward compatibility with AMPS networks. ANSI-136 allows the operators flexibility of deployment, i.e., the operators can choose to convert AMPS channels to ANSI-136 channels as the ANSI-136 traffic increases in the system. It is important to plan such deployment according to the traffic needs of the AMPS and ANSI-136 users present in the network.

Various authors have attempted different aspects of traffic analysis for cellular systems. A majority of these deal with traffic due to call origination and due to handovers [1], [2]. Mobile-to-mobile calls and PSTN-to-mobile calls are dealt with in [3]. The problems of dual-mode systems have not received much attention, one exception being [4].

In this chapter, we consider the problem of calculating the blocking probability for a partially overlaid AMPS/ANSI-136 cellular system, where some of the AMPS carriers have been replaced by ANSI-136 carriers each supporting three users. In this case, an approximation to the offered traffic for a certain blocking can be obtained by treating the two pools of channels as two independent systems and using the Erlang-B formula for each pool [4]. This approximation, however, is inexact due to the fact that ANSI-136 users will have dual-mode terminals, and will be admitted onto AMPS channels when ANSI-136 channels are unavailable. We derive the expression for the blocking probabilities for the two classes of users as a function of traffic for the case when dual-mode terminals are available. The system can be modeled as a two-dimensional Markov chain with a finite number of states and the blocking probability for the two classes of users can be derived using the steady state balance equations. The results also give insight into the percentage of AMPS carriers that need to be converted into ANSI-136 carriers to support a certain mix of traffic with a specified blocking probability.

We also propose an enhanced method, which controls the overflow of ANSI-136 users onto AMPS channels, and we find that an increase in the supported traffic can be obtained by such control. We derive a general framework that allows the calculation of the allowed traffic for different cases of overflow control into account, and derive strategies to increase the supported traffic.

We also consider the case wherein the transceivers at the base station can be configured quickly depending on the arriving traffic. Transceivers are nominally idle until they are required, and they are configured to support AMPS channels or ANSI-136 channels depending on the traffic needs.

Thus, a carrier normally used to support ANSI-136 may be converted to support AMPS if an AMPS user requests a channel, and no other free AMPS channel is not available. In this case, the derivation of the blocking probability is more involved. When intra-cell handovers are used to pack the ANSI-136 users, the problem is analytically tractable. The system can again be modeled as a two-dimensional Markov chain, and the blocking probability results can be derived.

When no packing of the ANSI-136 users is performed, many partially loaded ANSI-136 carriers may be found in the system. A carrier is released to be idle only if all the users on that carrier complete their calls. In this case, the analytical solution to the blocking probability is considerably involved and we do not attempt to perform the analysis. The blocking probability results, however, are obtained by means of a simulation. The results in this case are worse than the case when call packing is used due to the fact that channels are utilized less efficiently.

The chapter is organized as follows. In Section 2, we describe the analytical solution for the case of fixed number of carriers for AMPS and ANSI-136 and present some results. These results help motivate the discussion in Section 3, wherein we describe a paradigm in which the overflow of ANSI-136 users onto AMPS frequencies is controlled in order to increase the supported traffic. In Section 4.1, we consider the case of reconfigurable carriers with packing and perform the analysis. In Section 4.2, we describe the simulation for the case with reconfigurable carriers, but no packing. In Section 4.3, we consider the case of reconfigurable carriers with packing and controlled overflow. Analytical and simulation results are compared for the various cases. We conclude the chapter in Section 5.

2. FIXED PARTITIONING OF TRANSCEIVERS

With a fixed partitioning of AMPS and ANSI-136 transceivers, N transceivers (or N channels) are dedicated for AMPS and M channels (or $M/3$ transceivers) are dedicated to ANSI-136. An arriving AMPS call is blocked if all the N AMPS channels are occupied. If an arriving ANSI-136 call finds all M ANSI-136 channels blocked, it can still be assigned to an AMPS channel if it is available. Thus, an ANSI-136 call is blocked only if all AMPS and ANSI-36 channels are occupied.

A similar problem has been considered in the case of overflow systems in [2] and [5].

The state transition diagram of the system in terms of occupied AMPS and ANSI-136 channels is shown in Figure 1. The states are denoted $\{n, m\}$, where n is the number of active AMPS users and M is the number of active ANSI-136 users. An arrival rate of λ_a call/s is assumed for the AMPS users and an arrival rate of λ_d call/s is assumed for the ANSI-136 users. All arrivals are assumed Poisson. The holding time is assumed to be exponentially distributed with a mean of μ seconds. $\rho_a = \lambda_a/\mu$ and $\rho_d = \lambda_d/\mu$ are the normalized offered traffic values for AMPS and ANSI-136 users respectively.

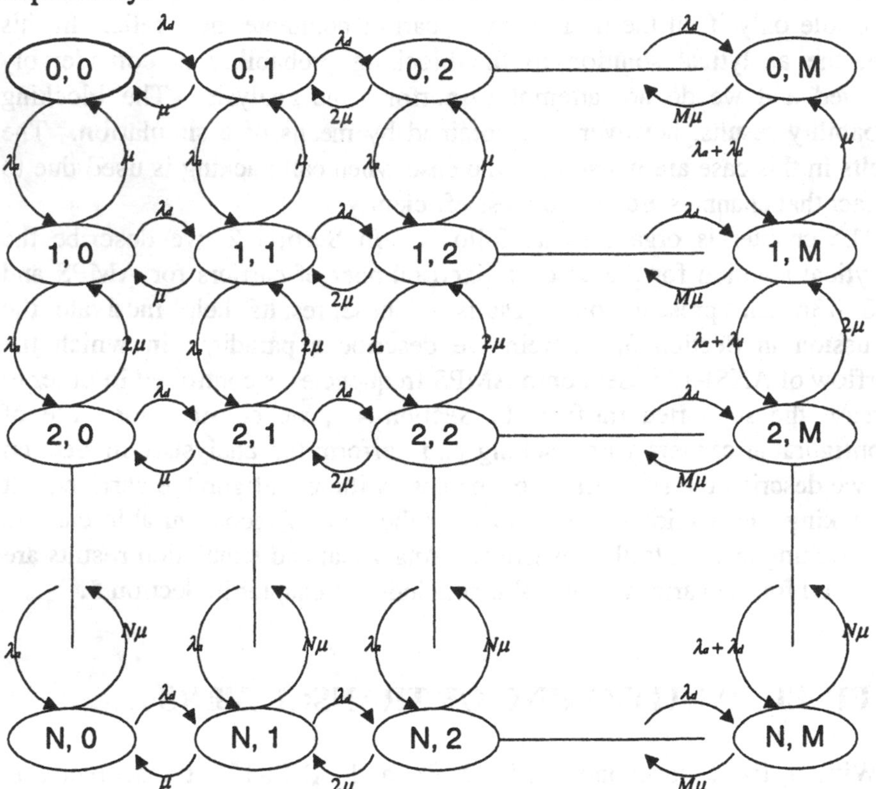

Figure 1. The state transition diagram for non-reconfigurable systems

From the figure, it is seen that:

1. Transitions between state $\{n,m\}$ and state $\{n,m + 1\}$ occur at a rate of λ_d.
2. Transitions between state $\{n,m\}$ and state $\{n + 1, m\}$ occur at a rate of λ_a.

3. Transitions between state $\{n,M\}$ and state $\{n+1, M\}$ occur at a rate of $\lambda_d + \lambda_a$ since all ANSI-136 channels are occupied and an AMPS or ANSI-136 call will be assigned to an empty AMPS channel.

Using the state transition diagram in Figure 1, we can solve for the stationary probabilities $P(n, m)$ of the various states $\{n, m\}$. Unfortunately, the structure of the diagram seems to be such that simplified solutions (e.g., a product form solution) do not appear possible. It can be noted that the state diagram is for an unbalanced system (due to the last column), and thus the general flow balance equations [6] do not hold. Thus, the solution has to be found by taking into account all possible state balance equations, and the normalization that all stationary state probabilities sum to unity.

The state balance equations are given by the following over-determined linear set:

$$
\begin{aligned}
&0 \le n \le N, \quad m = M: \\
&\quad (\rho_a + \rho_d + M + n)P(n, M) - \\
&\quad\quad (\rho_a + \rho_d)P(n-1, M) - \\
&\quad\quad\quad (n+1)P(n+1, M) - \\
&\quad\quad\quad\quad \rho_d P(n, M-1) = 0 \\
&0 \le n \le N, \quad 0 \le m < M: \\
&\quad (\rho_a + \rho_d + n + m)P(n, m) - \\
&\quad\quad \rho_a P(n-1, m) - \\
&\quad\quad\quad \rho_d P(n, m-1) - \\
&\quad\quad\quad\quad (n+1)P(n+1, m) - \\
&\quad\quad\quad\quad\quad (m+1)P(n, m+1) = 0 \\
&\text{and} \\
&\sum_{n=0}^{N} \sum_{m=0}^{M} P(n, m) = 1,
\end{aligned}
\tag{1}
$$

where all indices are bounded so that none of the flows are negative.

The quantities in which we are most interested are:

- The blocking probability for AMPS users P_a. This is given by

$$P_a = \sum_{m=0}^{M} P(N,m) \qquad (2)$$

- The blocking probability for ANSI-136 users P_d. This is given by

$$P_d = P(N,M) \qquad (3)$$

From the above equations, it is evident that $P_d \leq P_a$. Thus, the ANSI-136 users can always expect a better grade of service than the AMPS users. Using the above set of equations, we calculated the maximum number of ANSI-136 users that can be supported with a given amount of AMPS traffic that has to be supported with a certain grade of service. The mix of AMPS and ANSI-136 transceivers needed to support this maximum number of users was also found. The solution was found iteratively using an LMS based algorithm.

It is interesting to note that the problem of finding the global maximum traffic that can be supported with a system as described above is degenerate for any mix of M and N; the solution is that there must be no AMPS users and all ANSI-136 users accessing a total of $N+M$ channels.

2.1 Results and Discussion

We evaluated a system with 18 frequencies available for traffic. The two cases evaluated were:

- The pools of AMPS and ANSI-136 frequencies are independent
- If all ANSI-136 frequencies are in use, the ANSI-136 user can use an AMPS channel that is not in use.

For different AMPS traffic values, we calculated:

- The maximum allowable ANSI-136 traffic
- The mix of frequencies allocated to AMPS and ANSI-136 in order to support the calculated traffic values
- The actual blocking probabilities achieved

The supported ANSI-136 traffic for the two cases is shown in Figure 2. It is seen that a slight improvement in traffic is obtained with Case 2 (No reconfiguration) when the AMPS traffic that needs to be supported is high. As more and more ANSI-136 users use the network, however, the surprising result is that Case 2 is actually less efficient than the independent pool paradigm. Thus, it would be prudent for a service provider to allow ANSI-136 calls to overflow into AMPS channels under initial deployment, but as the digital network grows, it becomes worthwhile to treat ANSI-136 and AMPS channels independently.

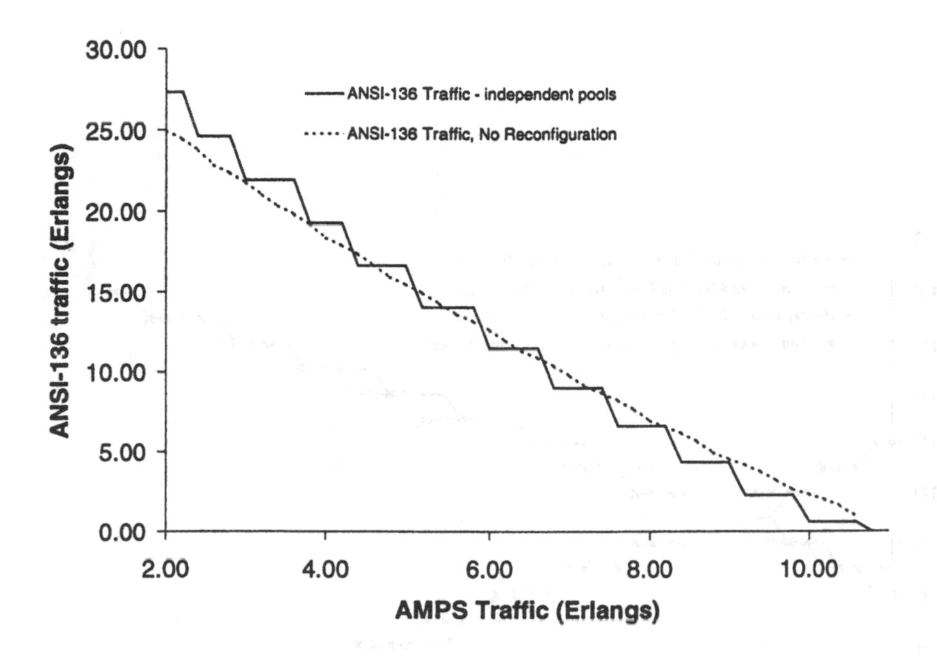

Figure 2. Supported traffic for the cases when AMPS and ANSI-136 carriers are in independent pools, and when overflow from ANSI-136 to AMPS is allowed.

The numbers of AMPS and ANSI-136 frequencies needed to achieve the maximum ANSI-136 traffic for a given AMPS traffic are shown in Figure 3. It is seen that the number of AMPS frequencies required is greater when overflow of ANSI-136 users is allowed. This is particularly true at low levels of AMPS traffic. This possibly explains the higher efficiency of the independent pool case at low AMPS traffic levels.

The actual blocking probabilities achieved for the two cases above for the AMPS and ANSI-136 users are shown in Figure 4. For the case of independent pools of frequencies, it is seen that the AMPS blocking probability is actually below the requirement of 2%. This is mainly due to the granularity of the number of trunks needed to support a given AMPS traffic. For this case, the blocking probability of ANSI-136 users is equal to 2%. In the case when ANSI-136 users overflow into AMPS, the AMPS blocking probability is increased to 2%, but the blocking probability of IS-136 users is extremely low. Thus, it is possible that there are schemes that control the overflow of ANSI-136 users onto AMPS, increase the blocking probability of ANSI-136 users up to the 2% level with more ANSI-136 traffic supported for a specified AMPS traffic. In the next section, we propose a general paradigm to look at such controlled overflow.

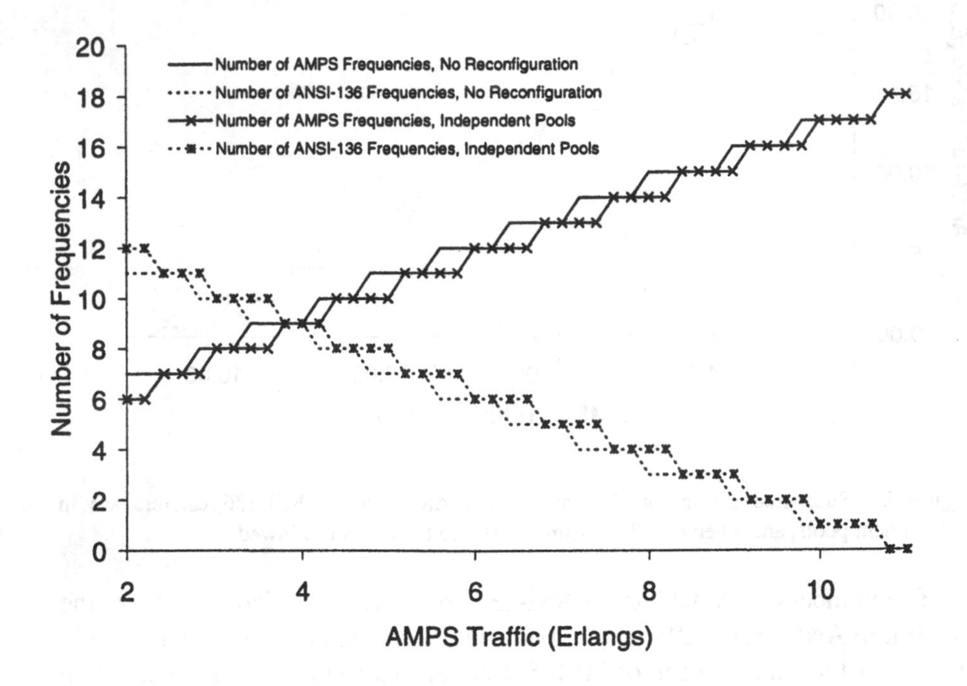

Figure 3. The number of frequencies assigned to AMPS and ANSI-136 when transceiver assignments are fixed for all time. The number of available digital channels is three times the number of frequencies.

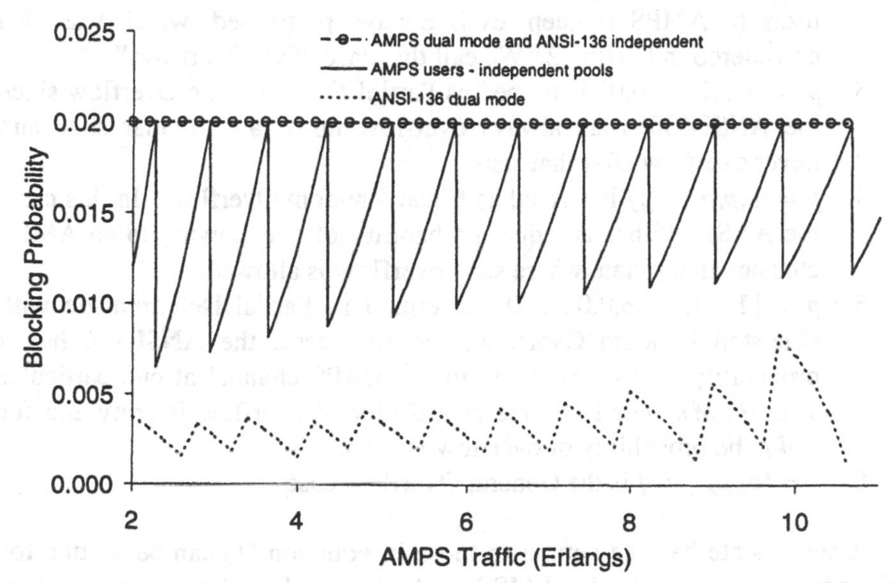

Figure 4. Actual blocking probabilities achieved for the cases of independentpools of channels and overflow from ANSI-136 to AMPS

3. CONTROLLED OVERFLOW PARADIGM

The overflow of ANSI-136 to AMPS frequencies can be controlled using probabilistic admission control. If all ANSI-136 channels are occupied, then the ANSI-136 user is allowed to overflow to an available AMPS frequency with a certain probability, which can be dependent on the number of AMPS frequencies available.

The state transition diagram of the system in terms of occupied AMPS and ANSI-136 channels is shown in Figure 5. This is similar to the state diagram in Figure 1, except for the states in the right column, where it is seen that the set of probabilities $\boldsymbol{p} = \{p_1, p_2, \ldots p_N\}$ modifies the arrival rate of the ANSI-136 calls when a transition to an AMPS frequency occurs. Thus, at each of the states (k, M), the probability that an ANSI-136 call will be assigned an AMPS frequency is equal to p_{k+1}. Many special cases can be derived using this paradigm for different assumptions on **p**. Some of these are enumerated below:

1. p = {0,0,...0} is equivalent to the case with independent pools of AMPS and ANSI-136 frequencies as given in Section 2.
2. p = {1,1,...1} is equivalent to the case where overflow of ANSI-136 users to AMPS frequencies is always performed, which was also considered in Section 2. We call this case "Full Overflow."
3. p = {1,1,1,... ,0,0} is termed as Partial Deterministic Overflow since the ANSI-136 users always overflow up to a particular state and never overflow after that state.
4. $p = \{p,p,p,\dots,p\}$ is termed as Equal Random Overflow. In this case, the ANSI-136 has an equal probability of overflowing to an AMPS channel at any state where such overflow is allowed.
5. p = {1,1,1,... ,p,0,0,... ,0} is termed as Partial Deterministic with One-step Random Overflow. In this case, the ANSI-136 has a probability p of overflowing to an AMPS channel at one particular state (k_1,M). For $k<k_1$, the probability of overflow is unity and for $k>k_1$, the probability of overflow is zero.
6. $p = \{p_1,p_2,\dots,p_N\}$ is the General Overflow case.

A set of state balance similar to those in equation (1) can be written for this case too, and solved. LMS-based search algorithms were used to optimize the value of the probability p for the Equal Random Overflow and the Partial Deterministic with One-Step Random Overflow cases. The results for the supported ANSI-136 traffic for a given AMPS traffic are shown in Figure 6. It is seen that the controlled overflow paradigm is able to outperform the independent pools case at all levels of AMPS traffic. Also, the best results are achieved with the Partial Deterministic with One-step Random Overflow case. However, the difference in supported traffic between this case and the Partial Deterministic Overflow case is rather small, thus the Partial Deterministic Overflow case might be preferable since the implementation is simpler.

For the Partial Deterministic Overflow Case, we show the number of AMPS frequencies needed and the allowable overflow AMPS channels in Figure 7.

A comparison with the AMPS frequencies needed for the Independent Pools Case and the Full Overflow Case shows that the number of frequencies needed for AMPS for the Partial Deterministic Overflow case is closer to that of the Independent Pools case. This is probably the reason why it is does not suffer from a loss of traffic when AMPS traffic is low. Also, the number of overflow channels is shown in Figure 6. The number of overflow channels shows some variation about a local mean which is around three lower than the number of AMPS frequencies in the system. Thus, it is conceivable that a practical system could allow overflow of ANSI-136 calls on to AMPS frequencies as long as there are more than three AMPS frequencies available, while blocking the ANSI-136 calls when there are less than 3 AMPS frequencies available. This strategy helps maximize the total traffic and provide adequate grade of service to both classes of users.

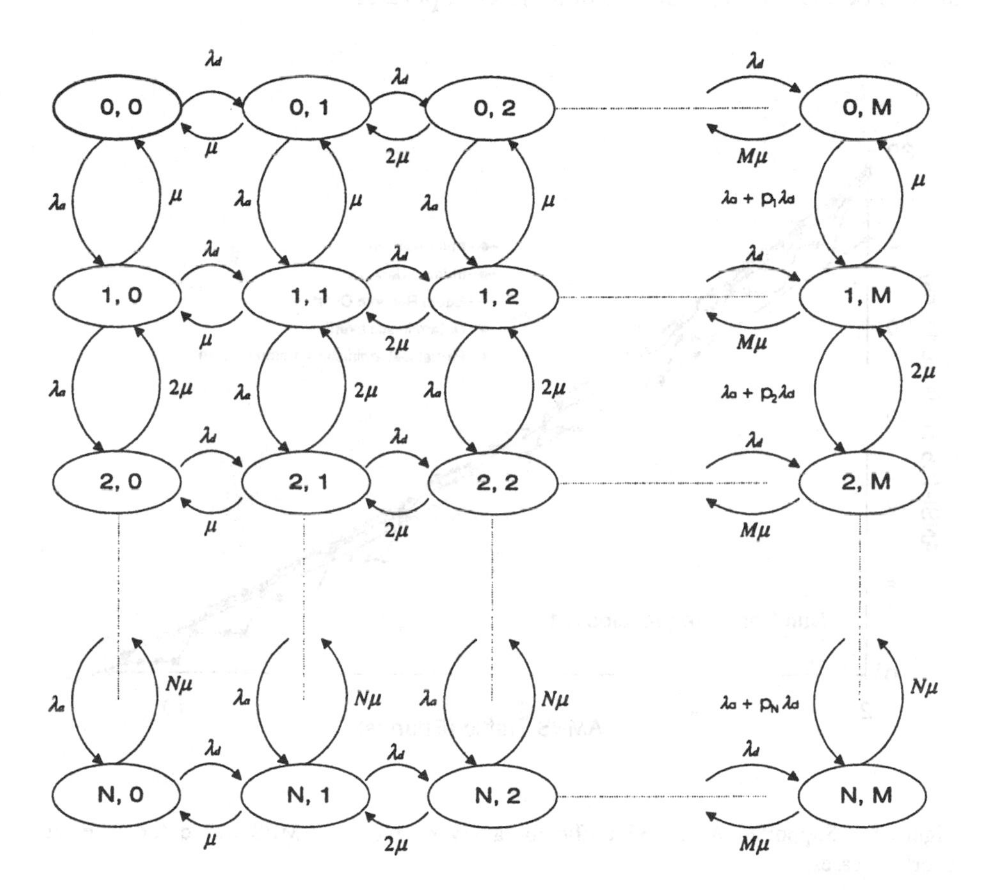

Figure 5. The State Diagram for Controlled Overflow of ANSI-136 users on to AMPS frequencies, where the probabilities $\{p_1, p_2, \ldots, p_N\}$ control the overflow

In Figure 8, we show the overflow probabilities for the Equal Random Overflow and Partial Deterministic with One-step Random Overflow cases. A large variation in the overflow probability is seen with varying AMPS traffic. For the Equal Random Overflow case, the general trend is an increase in the overflow probability for higher values of AMPS traffic, which indicates that the Full Overflow Case is optimum for large values of AMPS traffic. Nevertheless, it is difficult to optimize the overflow probability unless expected traffic values are precisely known. Thus, the Partial Deterministic Overflow method is preferable from an implementation viewpoint. Also, the Partial Deterministic Overflow method is better than the Equal Random Overflow method and only marginally worse than the Partial Deterministic with One-step Random Overflow method, thus it should be the preferred choice of a system operator.

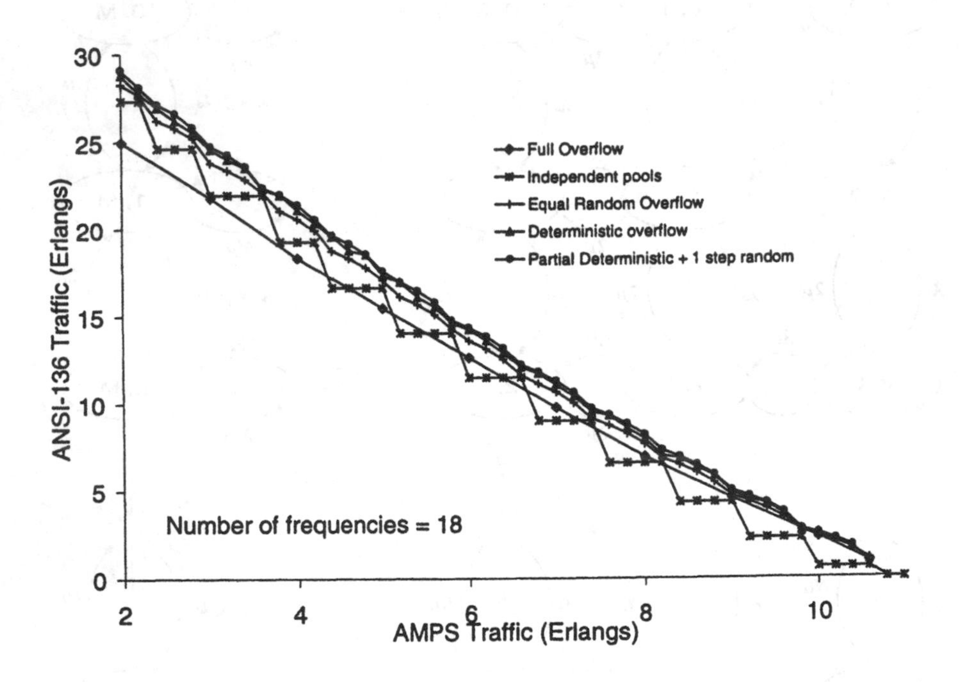

Figure 6. Supported ANSI-136 traffic for a given level of AMPS traffic for different overflow cases.

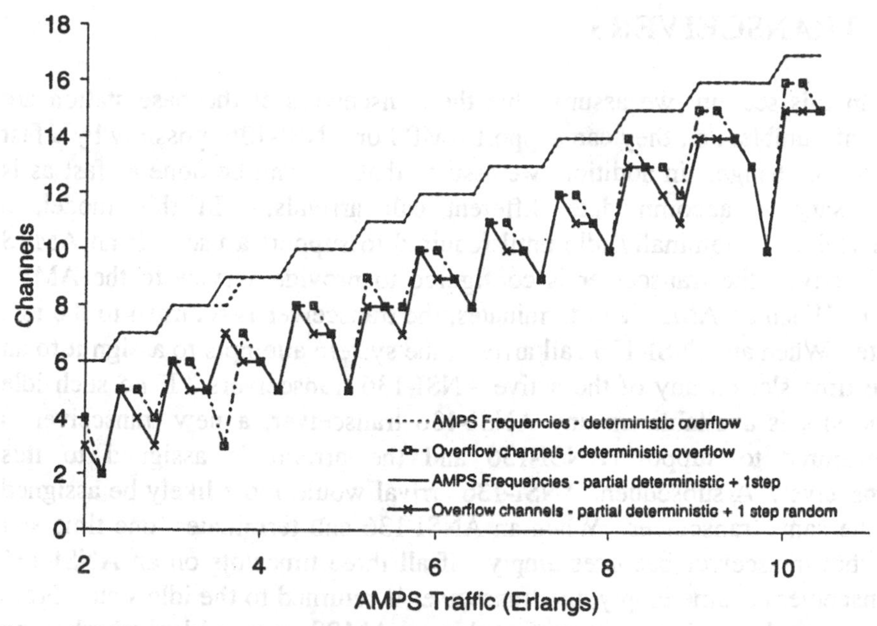

Figure 7. Number of AMPS frequencies needed and the number of overflow channels permitted for ANSI-136 for the Partial Deterministic Overflow case.

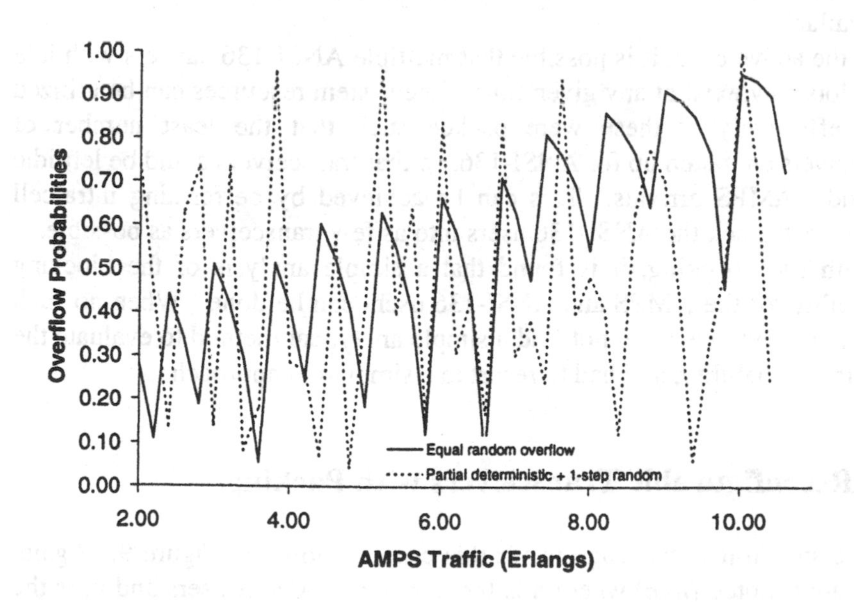

Figure 8. The Overflow Probabilities for the Equal Random Overflow and the Partial Deterministic with One-step Random Overflow cases.

4. RECONFIGURABLE BASE STATION TRANSCEIVERS

In this section, we assume that the transceivers at the base station are reconfigurable, i.e., they can support AMPS or ANSI-136, possibly by a fast software change. In addition, we assume that this can be done as fast as is necessary to accommodate different call arrivals. In this model, a transceiver is nominally idle until required to support a user. If an AMPS call arrives, the transceiver is configured to provide service to the AMPS user. When an AMPS call terminates, the transceiver is returned to the idle state. When an ANSI-136 call arrives, the system attempts to assign it to an idle time slot on any of the active ANSI-136 transceivers. If no such idle time slot is available on any ANSI-136 transceiver, a new transceiver is configured to support ANSI-136 and the arrival is assigned to this transceiver. A subsequent ANSI-136 arrival would most likely be assigned to the same transceiver. When an ANSI-136 call terminates, one time slot on that transceiver becomes empty. If all three timeslots on an ANSI-136 transceiver become empty, the transceiver is returned to the idle state. From the above discussion, it is evident that an AMPS user is blocked when no idle transceivers are available. An ANSI-136 user is blocked if no idle time slots on active ANSI-136 transceivers are available, and no idle transceivers are available.

In the above case, it is possible that multiple ANSI-136 carriers with idle time slots may exist at any given time. The system resources can be utilized more effectively if these were packed such that the least number of transceivers are taken up for ANSI-136, so that transceivers could be left idle to handle AMPS arrivals. This can be achieved by performing intra-cell handovers to pack the ANSI-136 users into as few transceivers as possible.

With such packing, it is found that a simple analysis of the blocking probability for the AMPS and ANSI-136 users can be done. When no such packing is used, we could not find a simple analytical method to evaluate the blocking probability, and had to resort to a simulation approach.

4.1 Reconfigurable Transceivers with Packing

The state transition diagram for this case is shown in Figure 9. Again, states are denoted (n,m) where n is the number of AMPS users and m is the number of ANSI-136 users. A mix of n AMPS users and N-n ANSI-136 users utilizes all transceivers. A departure of one AMPS user or three ANSI-136 users frees one transceiver, which can be configured to accommodate one arriving AMPS user or up to three arriving ANSI-136 users. The assumption made here is that intra-cell handovers are instantaneous.

The problem considered in [3], wherein PSTN-mobile calls take only one radio channel, whereas mobile-mobile calls take two radio channels has the same flavor as the one considered in this section.

It is observed that this state transition diagram is just a truncation of a state transition diagram of a system with two independent queues, one a *M/M/N* queue, and the other a *M/M/3N* queue. For such a truncation, it is known that a product form solution holds for the stationary state probabilities *P(m,n)* [6]. Thus, we have

$$P(m,n) = K \rho_a^n \rho_d^m \tag{4}$$

where K is a normalizing constant which is chosen to make the probabilities of all allowable states sum to unity.

The blocking probability P_a for AMPS users is given by

$$P_a = P(N,0) + \sum_{k=0}^{N-1} \sum_{l=1}^{3} P(k, 3(N = k - 1) + l). \tag{5}$$

The blocking probability P_d for ANSI-136 users is given by

$$P_d = \sum_{k=0}^{N} P(k, 3(N - k)) \tag{6}$$

It is seen that $P_d \leq Pa$, i.e., a better grade of service is available to the ANSI-136 users again.

Using the above equations, we were able to calculate the amount of ANSI-136 traffic that can be supported by the system while simultaneously supporting a given AMPS traffic with a certain grade of service. An iterative method was used to find the solution.

4.2 Transceivers without Intra-Cell Handover

In this case, multiple ANSI-136 transceivers may have idle channels and the analytical approach, if not intractable, is extremely cumbersome. Hence, we obtained the traffic results by means of a simulation. In the simulation, the transceivers are left idle and reconfigured as an AMPS transceiver or an ANSI-136 transceiver depending on the call arrivals. When all calls on an ANSI-136 transceiver or an AMPS transceiver (i.e., the one call) are completed, the transceiver is returned to the idle mode. Also, a new ANSI-136 call is assigned to an active ANSI-136 transceiver with idle slots, if it exists. Priority is given to ANSI-136 transceivers with two active time slots. A new ANSI-136 transceiver is set up only if all active ANSI-136 transceivers have no idle time slots. The simulation was even-based, and 10000 blocking events were simulated for each case.

Using an iterative method, we found the number of ANSI-136 users that can be supported by the system while supporting a given AMPS traffic with a certain grade of service.

4.3 Reconfigurable Transceivers with Packing and Probabilistic Assignment

The state transition diagram for this case is the analogue of that in Figure 5 for packing of channels and is shown in Figure 10. The last ANSI-136 carrier is assigned randomly with a probability drawn from the set $\{p_1, p_2, \ldots, p_{N-1}\}$. The distribution was studied for the case where the constant value for the probability of transition to the last digital carrier was used, i.e., $p_k = p$. A product form solution does not exist in this case, the state balance equations may be solved for the stationary distribution. Curiously, it is seen that for most AMPS traffic in the low and moderate ranges of interest, a probability of zero achieves the maximum possible ranges of interest, a probability of zero achieves the maximum possible ranges of interest, a probability of zero achieves the maximum possible ANSI-136 traffic. Since the capacity of an AMPS channel is lower than a digital carrier, the result is counter-intuitive. It is a simple matter to evaluate the optimizing value of probability p for AMPS traffic in the range 9.2-11 Erlangs.

The strange result suggests that a probabilistic assignment could be used for the last two or more ANSI-136 digital carriers, instead of the last. Such a solution may be of interest to the practicing engineer, and is left as an exercise for the reader.

4.4 Results and Discussion

As before, the results assume a single cell with eighteen frequencies available for traffic. A maximum blocking probability of 2% was allowed.

In Figure 11, we compare the amount of ANSI-136 traffic that can be supported in the network for the following four cases:

1. Independent pools
2. Full Overflow, i.e., ANSI-136 users overflow onto empty AMPS frequencies when no ANSI-136 channels are available
3. Reconfigurable transceivers with packaing and no intra-cell handover
4. Reconfigurable transceivers with packing and no intra-cell handover

It is seen that the case with reconfigurable channels offers considerable improvements in supported traffic only if intra-cell handovers are used to pack the users to free up as many AMPS channels as possible. With high AMPS traffic, reconfigurable transceivers help even without intra-cell handovers, but independent pools of channels seem to be better as the network evolves with more ANSI-136 users.

With call packing and intra-cell handover, a significant improvement in supported traffic is seen for all values of AMPS traffic. The complexity associated with this scheme, however, and the possible degradation in voice quality due to the handovers needs to be considered.

5. CONCLUSION

In this chapter, we have evaluated traffic aspects of partially deployed AMPS/ANSI-136 networks for various models of system flexibility by using a mix of analysis and simulation approaches. We find that a significant improvement in carried traffic can be obtained by using reconfigurable transceivers, and using intra-cell handovers to pack users. With a fixed partitioning of AMPS and ANSI-136 users, it is advantageous for the service provider to allow blocked ANSI-136 calls to overflow to AMPS channels at initial deployment, but there is a penalty if such overflow is allowed at higher levels of ANSI-136 traffic. Controlling the overflow helps to stem such loss in capacity.

Acknowledgment

We are grateful to Professor Arne Nilsson of North Carolina State University for his comments. We also thank Barbara Friedewald for typographic assistance.

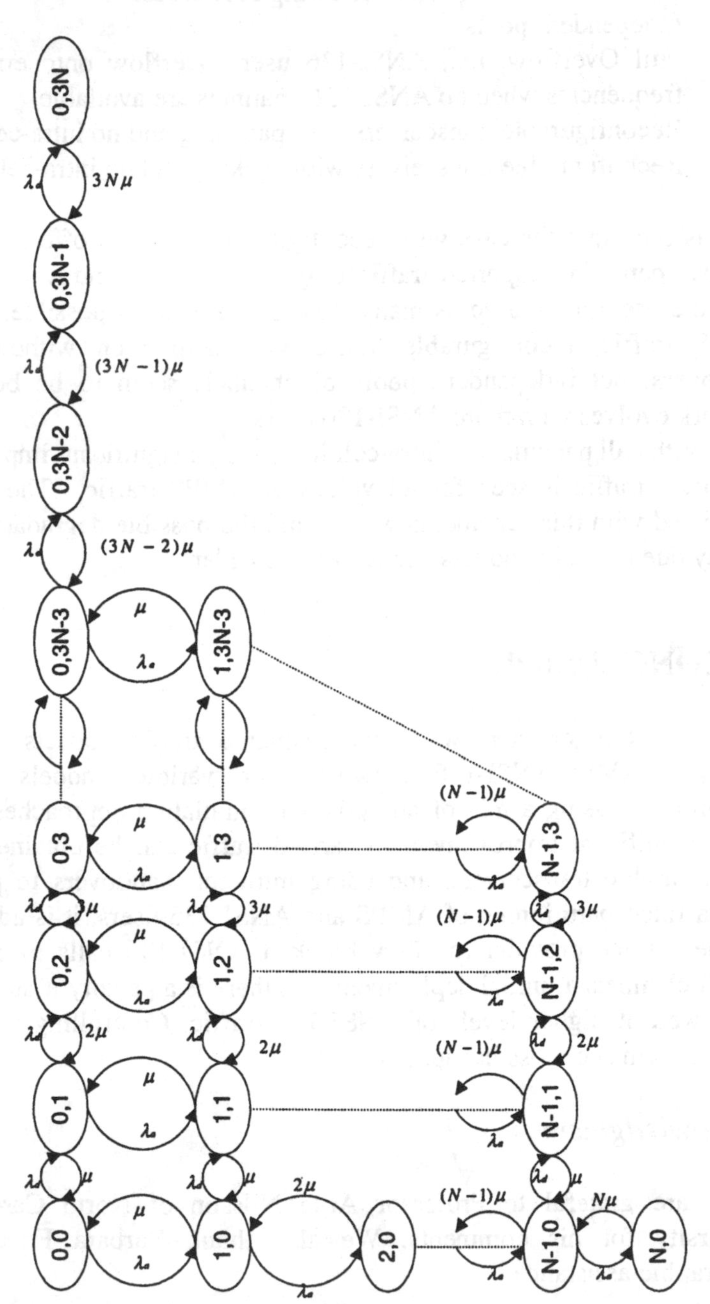

Figure 9. The state transition diagram for reconfigurable tranceivers with packing of ANSI-136 users.

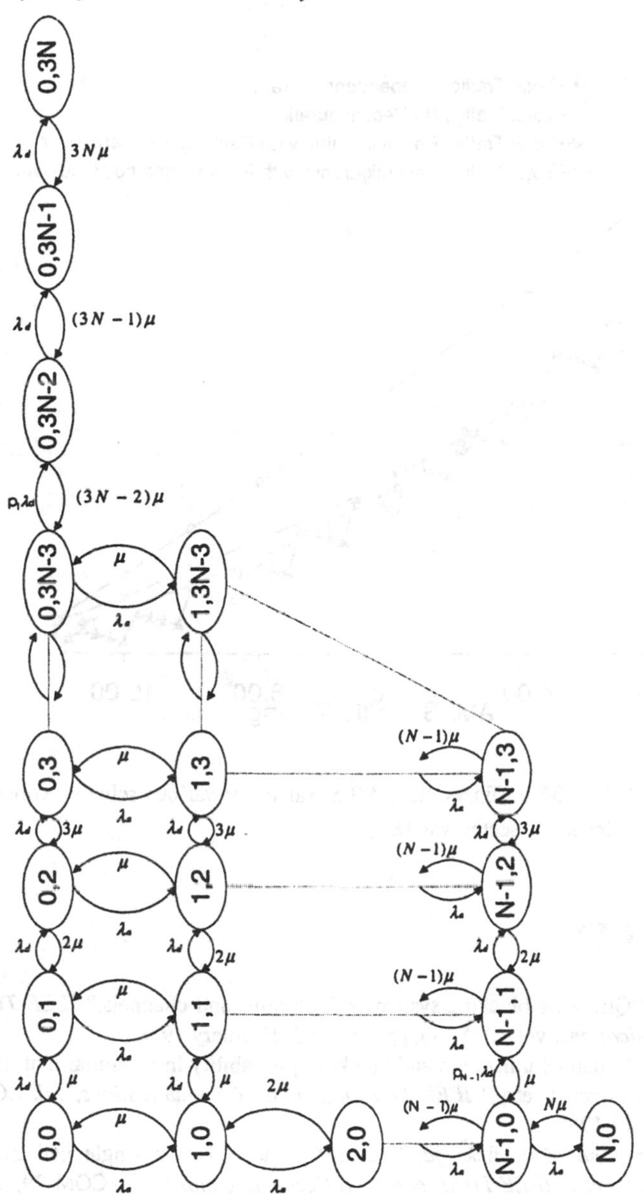

Figure 10. State transition diagram for deployment with packing of AMPS and ANSI-136 channels and probabilistic assignment for the last digital carrier

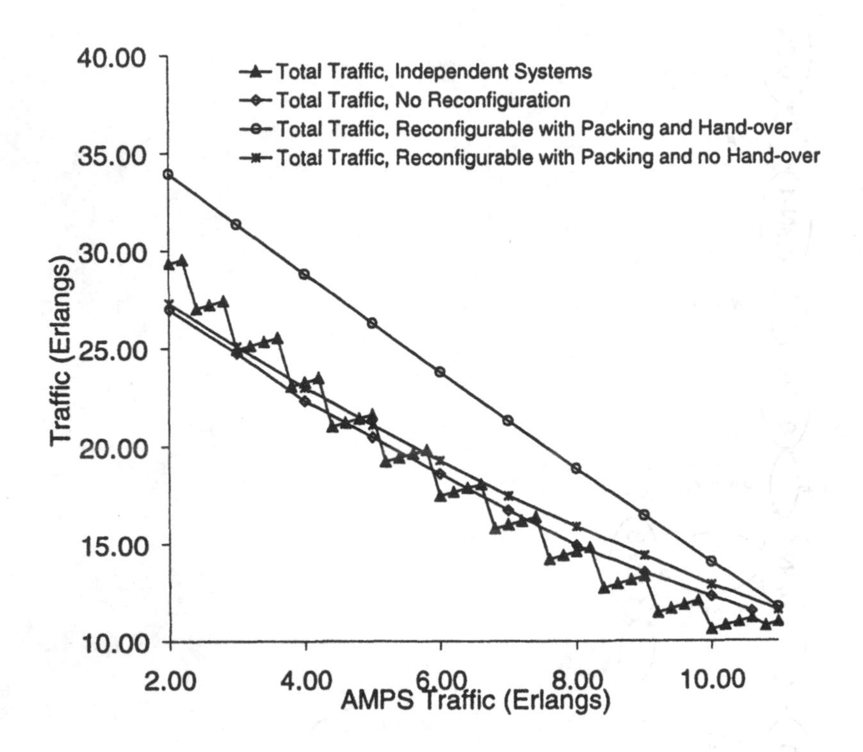

Figure 11. The ANSI-136 traffic vs. the AMPS traffic for various schemes considered. The number of frequencies in the sector was 18.

REFERENCES

1. *R. Guérin,* "Queueing-blocking system with streams and channels," *IEEE Transactions on Communications,* vol. COM-36, pp. 153-163, February 1988.
2. *B. Eklundh,* "Channel utilization and blocking probability in a cellular mobile telephone system with directed retry," *IEEE Transactions on Communications, vol.* COM-34, pp. 329-337, April 1986.
3. *S. H. Bakry and M.H. Ackroyd,* "Teletraffic analysis for single cell mobile radio telephone systems," *IEEE Transactions on Communications,* vol. COM-29, pp.298-304, March 1981.
4. *J. Shi et al,* "Throughput and trunking efficiency in the evolution of AMPS/DAMPS systems," in *International Conference on Universal Personal Communications,* vol. 2, pp. 864-867, Inst. Elect. Electron. Engr., 1997.
5. *R. B. Cooper,* Introduction to Queueing Theory, *ch. 4. Elsevier North Holland Ind., 1981.*
6. D. Bertsekas and R. Gallagher, Data Networks, ch.4. Englewood Cliffs, NH, USA: Prentice-Hall, 2nd ed., 1992.

Chapter 9

PRACTICAL DEPLOYMENT OF FREQUENCY HOPPING IN GSM NETWORKS FOR CAPACITY ENHANCEMENT

DR. ANWAR BAJWA

Camber Systemics Limited, UK

Abstract: GSM Frequency Hopping can realize increased capacity with marginal degradation in the Quality of Service. Measurements obtained from trial systems and operational networks have been presented to demonstrate that tight frequency reuse is viable. The behavior of the GSM RF parameters such as RXQUAL changes due to frequency hopping and the RF optimization of the cell parameters require careful attention. Synthesizer Frequency Hopping in a 1x1 frequency reuse pattern with 16% fractional load produces capacity gains in excess of 50%. It is also observed to be less sensitive to the variations in traffic load compared to the 1x3 fractional reuse. Fractional 1x3 reuse is more difficult to optimize in congested networks without traffic-directed congestion relief. Observations for cell parameter optimization are based on measured RF and network performance data from operational networks.

1. INTRODUCTION

Capacity expansion is a major planning consideration for all high growth networks. In the highly competitive cellular market network operators cannot afford to allow the network quality to deteriorate due to increasing congestion. In fact, an operator without a clearly defined capacity strategy is likely to respond with inappropriate and expensive solutions possibly leading to becoming even less competitive. GSM as a TDMA technology standard offers a number of basic features that enables the flexible deployment of increased capacity. Frequency Hopping is the most important wide area solution that has been successfully deployed by many GSM operators.

Although straightforward in concept, the practical deployment of Frequency Hopping is fraught with pitfalls. Firstly, each GSM vendor's product has evolved differently, ranging from proprietary algorithms implemented in the BSC software to engineering solutions that rely on standard cell parameters and optimization ingenuity. Embedded in some of the new features are complex system interactions such as traffic directed cell reselection and handover management. Secondly, a particular solution for one network does not always produce exactly similar results for another. The key considerations are the nature and distribution of the traffic load, the frequency hopping parameters, amount of spectrum and the cell architecture. The experience of optimising these networks will lead to advanced features and new optimization tools. Systematic measurements to improve network performance will allow intelligent algorithms to emerge.The current focus in GSM is shifting to the issues surrounding the introduction of high-speed General Packet Radio System (GPRS). This will require a new set of Quality of Service metrics for network planning and optimization.

The chapter will focus on a review of GSM frequency hopping schemes, the basic principles, the planning parameters and the practical aspects of implementation. The concept of capacity and quality is considered with reference to the interference averaging caused by frequency hopping. This is linked to the soft blocking in the context of Quality of Service (QoS). As there are a number of frequency reuse schemes that have been implemented a brief description of the main types is presented next. The practical performance data and results are limited to fractional Synthesizer Frequency Hopping (SFH) as this implementation has shown the greatest promise in many trial and operational networks. Finally the experience of deploying such networks has been summarised with suitable comments based on the results obtained from a number of GSM networks.

GSM Frequency Hopping is presented here as a first step in the capacity strategy. This is generally true and although microcells, indoor pico cells and dual band solutions are alternatives the capacity enhancement with frequency hopping is obtained with greater cost-effectiveness.

1.1 GSM Frequency Hopping

The physical TDMA time slot channels in GSM are arranged for transmissions on a fixed carrier frequency only for the duration of each burst. The carrier frequency is allowed to change, when the next burst arrives, for the same time slot in the next frame. This slow frequency hopping at approximately 217 hops per second provides a means to simultaneously realize frequency diversity and interference diversity, depending on the carrier frequency set available in each cell and the carrier hop distance.

The block-interleaving scheme in GSM spreads the bits in each coded speech block of 456 bits over 8 speech sub-blocks. Diagonal interleaving disperses bits from different time slots over successive bursts [1]. This further randomizes the bits from each speech block and with the channel coding, it decreases the likelihood of corrupting a complete speech frame due to a deep signal fade or interference. However a stationary mobile in a deep fade or persistently encountering co-channel interference will obtain little relief from even this depth of interleaving and powerful coding.

Allowing the carrier to hop, the channel frequency selectivity or the continuous co-channel interference becomes statistically time-dependent. The interleaving starts to work again by dispersing the error bits for correction by the channel decoder. In effect this process makes channels experiencing very poor conditions improve statistically to a level approaching an 'average' performance.

This simplified explanation introduces the basic concept, while the actual behavior is more complex and the dynamic modeling in simulations or a combination of analytical treatment and simulation is the only way to estimate the performance and capacity [2-4]. Though these analyses are theoretically interesting, they do not fully model the system behavior especially for scenarios with non-uniform and dynamically changing traffic load. They are all aimed at calculating capacity bounds in idealized homogeneous network scenarios. Capacity concepts adopted for CDMA systems are helpful for gaining some insight but since the channel and system management in GSM is different, the details do not translate directly and the analyses are not accurate [4].

To gain an appreciation of the practical aspects, it is necessary to consider the implementation within the context of field trials. This approach has to be carefully planned as the system interaction and the nature of some of the features e.g. traffic directed handover can cause second-order effects.

As most of these features are proprietary to the supplier of the system the results can be biased by the actual feature implementation. The best way to introduce these and related topics is to first look at generic FH planning and identify the key parameters for planning the system.

Frequency Hopping Parameters

In applying the basic rules for FH planning, it is important to first understand the parameters that influence the characteristics of Frequency Hopping systems. These parameters must be combined with a thorough understanding of the standard planning requirements e.g. traffic engineering.

Basic parameters

Frequency Hopping (FH) parameters are set in the system database and sent as general parameters for all mobiles in a cell and separately as specific parameters at call set up or handover.

Every hopping mobile must hop according to the hopping sequence set by the base station, the hopping sequence is derived by the mobile from general parameters sent regularly in the Broadcast Control Channel (BCCH) and the Synchronisation Channel (SCH). The general parameters are:

- CA or Cell Allocation as the set of carrier frequencies that are allowed for use at the particular base station that sent this information on the BCCH.
- FN or Frame Number of the current TDMA frame broadcast as a set of three parameters on the SCH.

The connection-oriented specific parameters sent as part of the channel assignment message are:

- MA or Mobile Allocation as the set of carrier frequencies i.e. a list to be used by the mobile in the hopping sequence.
- MAIO or the MA Index Offset as a parameter that ensures that mobiles using the same TDMA time slot number and the same MA list always hop to different frequencies. Because of this role MAIO is one of the important planning parameters.
- HSN or Hopping Sequence Number enables the selection of each successive frequency according to a pseudo random sequence based algorithm. Basically cells that use the same MA list and employ random hopping should always be assigned different HSN.

1.2 Capacity and Performance

The ideal capacity strategy would be to design and plan a network rollout that exactly matches the offered traffic to subscriber demand and remains synchronized with growth in demand over time. This strategy of network investment, if at all possible, would achieve optimum cost effectiveness. In practice the network is planned for flexible capacity expansion with some provisioning ahead of demand. This approach takes into account factors such as site acquisition, constraints or delays, and the uncertainty of the traffic growth both in terms of volume and the location of the demand.

Capacity can be increased by deploying small outdoor cells and by extending indoor coverage with dedicated indoor systems. Coverage improves with the deployment of small cells but the infrastructure investment becomes disproportionately large if the traffic hotspots are not accurately located. The cost of the sites, equipment and backhaul transmission typically accounts for 70% of the capital investment. Wide area deployment of micro-cells involves considerable investment even when confined to selected business districts in dense urban areas. At 900MHz estimates show between 25 to 40 micro-cells per square km. To selectively deploy outdoor micro-cells e.g. without a reliable indication of the high traffic local areas i.e. hotspots, can be risky and therefore it is not recommended as the first stage of an area-wide capacity strategy. Even if the traffic hotspot were known, the delivery of capacity requires a good frequency plan for the larger macro-cells to co-exist with the smaller micro-cells with minimum mutual interference. This means that a micro-cell must capture and handle traffic in its own coverage area at all times. It is possible to do this in the layered-cell architecture by dedicating a small portion of the spectrum for micro-cells. The right balance in the spectrum allocation is dependent on estimating the traffic levels accurately. Increased congestion could result in the outdoor macrocells if the correct balance in the spectrum allocation is not achieved.

Frequency Hopping is flexible and allows for increased capacity with marginal Quality of Service (QoS) degradation. To explain this mechanism the concept of soft blocking must be understood. By choosing the QoS performance threshold the traffic load handling can be varied between the soft blocking and hard blocking limits as depicted in Figure 1. The hard blocking limit is determined by the Grade of Service (GoS) i.e., call blocking for number of available channels per cell. Soft blocking to a certain extent depends on the traffic demand distribution as interference is generated by traffic in other cells. It can be managed by monitoring parameters such as Bit Error Rate(BER), Frame Erasure Rate (FER) and the Carrier to Interference Ratio (C/I) to set the thresholds for decisions that alter the system behavior.

The GoS and the Dropped Call Ratio determine the system QoS. The Dropped Call Ratio depends on the planning and practical implementation of the frequency reuse, it typically varies with the traffic load and congestion levels. The maximum capacity is defined at the crossover point between soft and hard capacity limits. The traffic carried in a cell then becomes strongly dependent on the number of available channels in the serving cell with limited opportunity for re-balancing call quality to increase capacity.

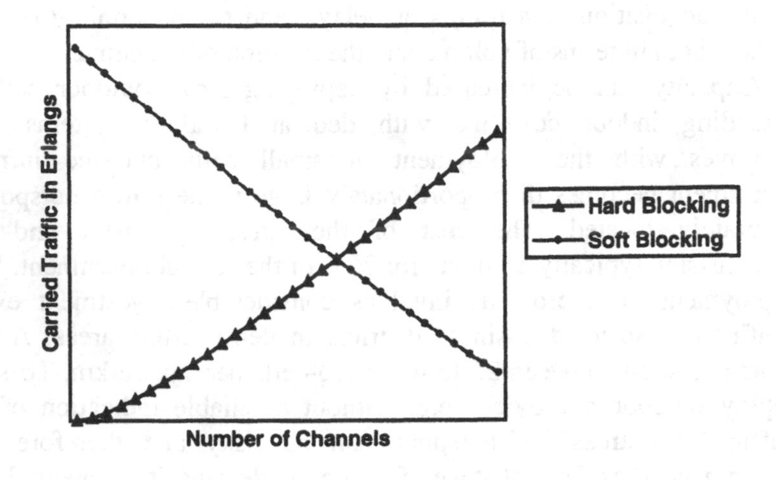

Figure 1. Capacity and QoS characteristics with soft blocking

The physical constraints of accurately measuring and reporting parameters limit the monitoring of soft blocking. In practical terms there are four parameters that can be observed and effectively used in current GSM systems for RF planning and RF optimization. The four possible candidate indicators for soft blocking are:

- RXQUAL
- C/I
- FER
- Dropped Call Ratio

RXQUAL and C/I measurements on the idle time slots at the BTS are standard measurement reports that are used in triggering Power Control (PC) and Handover (HO) for quality reasons. RXQUAL is a raw BER indicator and unlike the FER, does not always correlate with voice quality. FER is a better indicator but it is not currently supported in GSM measurement reporting. However it can be monitored to calibrate for voice quality and used for indirectly adjusting cell parameters. The Dropped Call Ratio is an counter available from the Operations and Maintenance (OMC) for off-line processing of statistics. The Dropped Call Ratio has been traditionally used in the performance monitoring and optimization of cellular systems. This indicator is also closely linked to the Radio Link Time-out (RLT) which is determined by the decoding failure rate of the SACCH frames. Although widely used, the indicator only indirectly represents the performance of the Traffic Channel (TCH). Therefore in certain frequency reuse scenarios, it cannot always provide accurate indication of the TCH quality.

Both RXQUAL and FER can be measured simultaneously with Test Mobile equipment and at the BTS with A-bis Call Trace measurement facilities. These are special arrangements that are needed in the optimization stages because the behavior of RXQUAL with Frequency Hopping is different to non-hopped systems. One way to show this is to plot the system reported Dropped Call Ratio against the number of events where the RXQUAL exceeds a threshold level e.g. RXQUAL greater than 5 in a cell. This gives an area-wide impression of the call quality, which involves many mobiles and reflects the true behavior for the RXQUAL parameter. The cell parameters in GSM are defined on a per cell basis and the RF optimization is performed by adjusting the thresholds for these parameters in terms of the reported parameters e.g. RXQUAL and RXLEV. The drive tests are useful to build a detailed log of the behavior in known problem areas. The plot in Figure 2 shows that the Dropped Call Ratio against the percentage of bad quality of calls, defined as the events where RXQUAL exceeds 5. The observed data confirms that the Dropped Call Ratio does not have a strong dependence on bad quality defined by the RXQUAL threshold. This behavior is due to the averaging effects of interference in Frequency Hopping systems.

Interference Averaging

Carrier frequency hopping causes interference from close-in and far-off mobiles to change with each hop. This means that a mobile continually suffering severe interference in a non-hopped case would be expected to experience lower interference due to the statistical averaging effect. The significance of this effect expressed in a simplified way translates to:

- The average interference during a call is lower and the average call quality is improved.
- The standard deviation of the interference is expected to become less, as the extreme events are fewer per call. For the same C/I outage the interference margin is reduced resulting in a lower C/I threshold.

This lower C/I cannot be directly mapped into a planning threshold. A determination of the quality threshold in terms of Frame Erasure Rate (FER) is a prerequisite as it is directly related to voice quality. This means that standard planning tools do not accurately reflect practical network quality and the frequency plans produced cannot be depended upon to evaluate capacity.

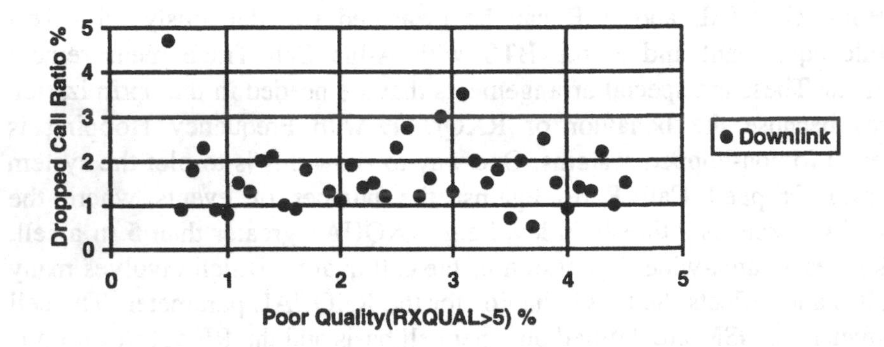

Figure 2. Soft blocking measurements: Dropped Call Ratio dependence on RXQUAL

Voice Quality and FER

The quality gain is not directly related to the mean C/I. This is because a certain mean C/I can result in different Frame Erasure Rates (FER) and unlike the non-hopped case where there is a unique mapping between the two parameters. The interference averaging causes the C/I distribution to change in a way that short term C/I are individually related to each FER, and the mean C/I can be identified with more than one FER distribution. This relationship has been observed in detailed system simulations based on snap-shot locations of mobiles over a large area and by assuming different traffic intensity per mobile. A sample result from simulations based on a homogeneous network of 50 sites covering an area of approximately 1500 square km, uniform offered traffic intensity of 25mE per mobile and spectrum allocation of 36 carriers is shown in Figure 3. The effects of downlink power control and Discontinuous Transmission (DTX) were modeled in these simulations.

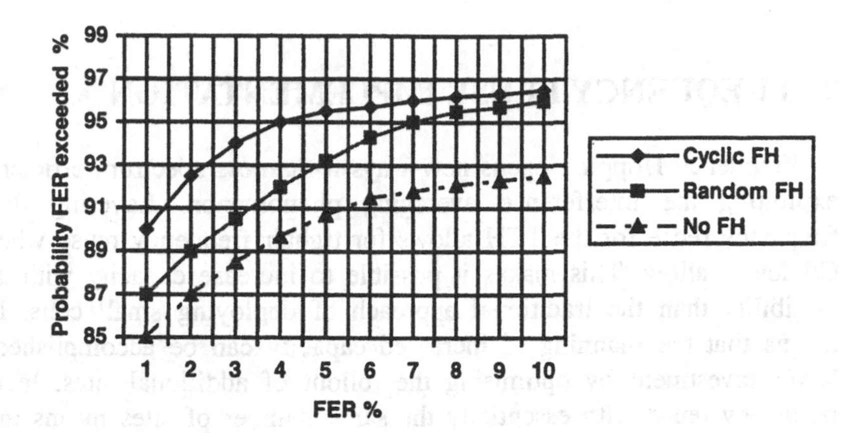

Figure 3. System Performance simulations: FER in an homogeneous network

In practical tests the FER and voice quality improve even though the mean C/I threshold is lowered. The plot of cumulative probability of the FER shows that with FH the 2% FER level is exceeded in 90 % of locations over the coverage area. At this FER level good speech quality is generally obtained in GSM systems. The better performance for cyclic FH is a manifestation of the channel modeling in the simulation and should not be interpreted as a superior gain compared to random FH. Uncorrelated TDMA bursts were simulated that produced maximum gain for cyclic FH.

Frequency planning for Frequency Hopped systems is not based on the worst case C/I as interference averaging alters the C/I statistics, instead the threshold C/I is adjusted to a lower mean value. A tighter frequency reuse is achieved in this way. This potentially effectively creates the potential for extra capacity. Capacity realized in this way can be exploited to either reduce congestion or enhance the call quality over a wide area. The improved system performance has been observed in many trials as well as operational networks.

Power Control and DTX

Power control at the BTS in conjunction with DTX can be used to reduce the level of interference. The activation of DTX creates transmission pauses during the silent periods in the speech. The BTS has a limited range for power control but even allowing for this there can be significant gains in activating this, in association with DTX to achieve interference reduction.

The gain from these features can be exploited usefully to achieve better quality with tight frequency reuse.

2. FREQUENCY REUSE IMPLEMENTATION

Frequency Hopping opens new ways to harness spectrum efficiency by exploiting the interference averaging phenomenon. Layering different frequency reuse for the TCH allows for tighter frequency reuse where the C/I levels allow. This makes it possible to increase capacity with greater flexibility than the traditional approach of deploying small cells. It also means that the planning of increased capacity can be accomplished with lower investment by optimising the rollout of additional sites. Increased frequency reuse with essentially the same number of sites means that the first stage of capacity expansion can focus on adding more equipment in the form of TRX and BTS rather than new sites for capacity expansion. This has a major benefit for network operators in optimising the network rollout investment. Even allowing for some additional sites for traffic hotspots e.g. micro-cell or indoor cells, this forms the basis of a cost-effective capacity strategy.

In practical systems the BCCH frequency plan is treated as a separate layer as in most implementations, the BCCH carrier is not allowed to hop. Therefore the reuse chosen for this layer is conservative compared to the TCH. Most networks deploy a frequency reuse equivalent to 4x3 i.e. four site x 3 cell repeat pattern.

Novel implementations have evolved with each of the major infrastructure equipment vendors offering features based on three generic schemes:

- Multiple Reuse Patterns (MRP) or layered frequency plan
- Intelligent Underlay-Overlay (IUO) or Intelligent Layered Reuse
- Fractional Load Reuse with Synthesizer Frequency Hopping (FL-SFH)

Multiple Reuse Pattern

Multiple Reuse Pattern is a layered frequency reuse scheme in which TCH carriers, arranged in frequency groups for each layer, are planned with a different reuse pattern. One layer may be planned with tighter reuse compared to another layer. This is possible because the traditional frequency reuse planning is typically based on the worst case C/I threshold, and on average the C/I requirement can be relaxed if the aggregate interference is lower. The C/I margin can be sacrificed in return for greater capacity without perceptible loss of quality. In all cases the BCCH frequency reuse is maintained in a separate high quality layer.

Multiple Reuse Pattern can be deployed without the need for a new software feature in the BSC. It can be planned with standard planning tools with some attention to the interference thresholds but to achieve good results, it is usual to give particular attention to HSN and MAIO assignment. The planning and implementation essentially form a part of an engineering solution that requires BTS hardware and database reconfiguration i.e. each TRX is identified to a frequency sub-group of the TCH layer. Hardware changes depend on the band segmentation and the type of transmitter combiner used.

The main drawback of Multiple Reuse Pattern is the reduction in spectrum utilization efficiency due to the reduced trunking gain i.e. fewer frequencies per sub-group especially where the actual traffic load is not matched to the layered reuse in the particular area, effectively causing a reduction in carrier utilization. This deficiency has been overcome in some networks in a novel way by combining MRP with Fractional Load, also known as FL-MRP.

Intelligent Underlay -Overlay

The original concept was proposed as a cost-efficient capacity expansion solution by introducing dual-layer channel segmentation on existing sites in an area of high demand. The concept is based on the assumption that mobiles close to the BTS site in general will have better C/I. Therefore a tight reuse (super-reuse) could be planned for a smaller concentric zone around the BTS site. The BSC dynamically calculates the C/I and assigns a mobile to a channel in the super-reuse or regular reuse layer by performing an inter-layer intra-cell handover.

The IUO algorithm has to be implemented in the BSC software and activated in the selected areas. This involves a modification of the system databases and TRX reconfiguration. The C/I assessment in the IUO algorithm is based on signal strength measurements of the BCCH carriers of the neighboring cells. The downlink measurements are done by the mobile in the idle time slots and reported back. The uplink measurements are also available to the BSC for overall C/I estimation and handover decision making.

The traffic absorption in the super reuse layer is known to be sensitive to the traffic distribution i.e. how much of the traffic demand is in close proximity to the BTS site. If the geographical traffic load distribution is not concentrated nearer the BTS site locations the traffic absorption is not high.

Fractional Reuse

Fractional reuse minimises the probability of carrier collisions, hopping over a set of frequencies greater than the actual TRX deployed in each cell. The co-channel or adjacent interference caused by collisions or hits of the hopping carriers depends on the ratio of the number of TRX and the frequency allocation of the reuse group. The lower the ratio the lower the probability of a carrier hit and therefore this ratio is termed the Fractional Loading (FL), meaning the fraction of frequencies that actually transmit at any time. Fractional loading is correlated with the traffic load and the performance of the high capacity solution depends on the basic traffic demand characteristics.

Fractional loading of the carriers is possible by using Synthesizer FH (SFH) since the carrier must hop to a different frequency over a larger set of frequencies from one GSM TDMA time frame to the next. GSM does not allow time slots to be changed for dedicated TCH channels without the involvement of a handover.

This solution assumes that the fractional loading can be planned in a way to match the actual traffic demand. Choosing the tightest frequency reuse increases the available carrier set in each cell, and therefore potentially enables operation at a lower Fractional Load (FL). FL-SFH with 1x1 reuse i.e. N=1 reuse, generally has reduced sensitivity to the dynamics of traffic load compared to 1x3 FL-SFH with N=3 reuse and can deliver higher overall capacity. However the practical solutions require careful attention to the MAIO and HSN parameter planning, especially for adjacent channel interference control.

3. PERFORMANCE OF PRACTICAL NETWORKS

Baseband Frequency Hopping first demonstrated the practical performance gain from Frequency Hopping. The implementation at the time was limited to specific MRP and IUO deployment in certain mature GSM networks. Aggressive frequency reuse schemes based on the 1x3 and 1x3 fractional SFH since then have been successfully tested in many pilot trials, and recently a number of network operators have deployed these reuse schemes in operational networks. Early experience has been encouraging and results suggest that there is significant potential for capacity enhancement with SFH. The use of downlink power control and DTX have generally produced better results, but the comparative data for the same network is limited to selected measurements from trial networks. There are also some reports of successfully combining other traffic-directed features for umbrella cell, underlay-overlay and concentric cell deployment scenarios. This evaluation is currently in progress in different operational settings of high capacity networks.

3.1 Fractional SFH Network Performance

The performance of fractional SFH systems has been presented to demonstrate the relevance of the practical results and to establish the basic relationships between the parameters of interest. There is a combination of data from pilot trials and also from selected networks. The available data and the form of the data is limited because of commercial sensitivity. However there is sufficient consistency in the results which allows for key observations and verification of the main claims.

Scenarios and Objectives

In fractional reuse the major parameter that influences the capacity is the fractional load. In a number of pilot trials the scenarios were deliberately arranged to study the characteristics of fractional load. Fractional load can be changed in situations where there is sufficient flexibility to increase the frequencies in the MA list or to modify the TRX configuration in each cell. This has to be done with reference to the traffic load and the congestion or GoS level in a given area. In some cases the load was adjusted by removing TRX in a cell to establish the operating point for the traffic load and to study the sensitivity of the frequency reuse to traffic load variations. The QoS level variations and the soft blocking characteristics were also studied in this case. This approach was adopted, as the trial networks were limited to observations over a few weeks during which the volume of traffic was not expected to increase dramatically. Data from operational SFH networks is accumulating over time but limited to a specific scenario in the area and highly dependent on the extent of RF optimization performed.

The typical trial system involved 20 to 30 sites over an area less than 10kmx10km. In the networks considered here down link power control and DTX were activated with frequency hopping.

The interesting scenarios from an implementation perspective included:

- Reference system with typically a 4x3 frequency reuse
- 1x3 SFH fractional reuse with 25, 33 and possibly 50% fractional load
- 1x1 SFH fractional reuse with 8 and 16% fractional load

Although the objectives of the trials varied depending on the network operator's main priority the main objectives were:

- Estimation of the capacity gain for the allocated spectrum
- Verify that the voice quality or QoS requirement is met in the worst case
- Understand the sensitivity of the tight frequency reuse to practical planning and deployment

The experience of the trial networks has helped many operators to refine the parameters for operational conditions. This involved extensive optimization activities, especially to ensure that the interactions between features were understood prior to the launch of a wide area network. The operational network for some operators has served as the live validation network for implementing aggressive fractional frequency reuse in a layered network architecture with micro and pico-cells.

Performance Statistics

The performance evaluation looks at the RF performance in terms of the radio parameters and Network or System performance in terms of the analysis of OMC counters. The RF performance statistics presented include the BER behavior as a function of RXQUAL and RXLEV and voice quality in terms of subjective and objective tests.

Impact of Fractional load

RXQUAL is a raw BER indicator, and the characteristics evident in Figure 4 suggest that power control and handover thresholds based on RXQUAL would cause an increased incidence in the triggering of such events. The percentage of RXQUAL samples relative to the traditional 4x3 frequency reuse can be more than four times greater. The increasing fractional load also causes a peaking of values around levels 4 and 5.

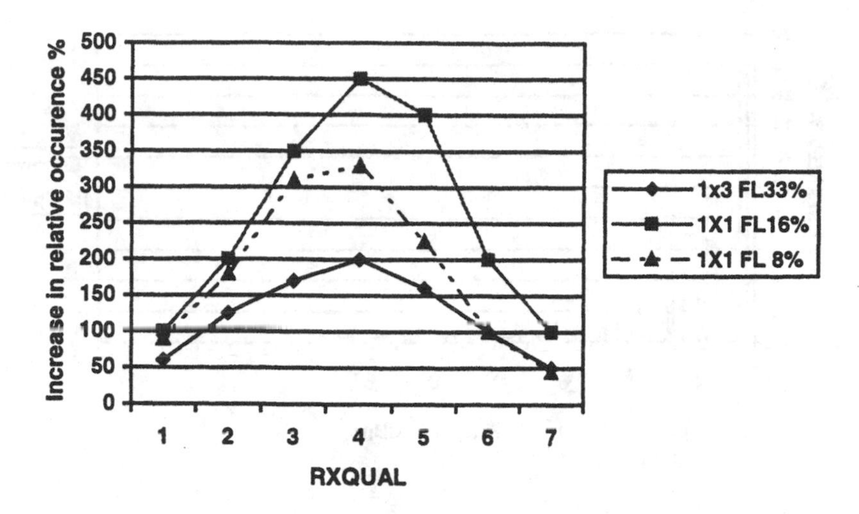

Figure 4. Impact of fractional load on RXQUAL

Both the Power Budget (PBGT) and RXLEV based handovers are triggered by RXLEV threshold and it is important to understand the interaction between the raw BER and RXLEV. The data for a 1x1 SFH system with 16% fractional load is shown in Figure 5 give a useful indication of the expected average BER within the operating RXLEV window after the first stage of RF optimization. The upper and lower thresholds can be also estimated from such data for RXQUAL for setting the power control window.

Voice quality and RXQUAL

Subjective voice quality assessment involves informal listening or conversational tests. To arrange formal tests is very time consuming and in most cases the tests are performed by equipment that estimates the Mean Opinion Score or an Audio Test mean from the sampled data. The important step in the analysis is to relate the samples of the Audio Test mean obtained over a suitable period for each RXQUAL level. Although the FER is a better indicator of speech quality, it is not available as a parameter for setting the trigger thresholds.

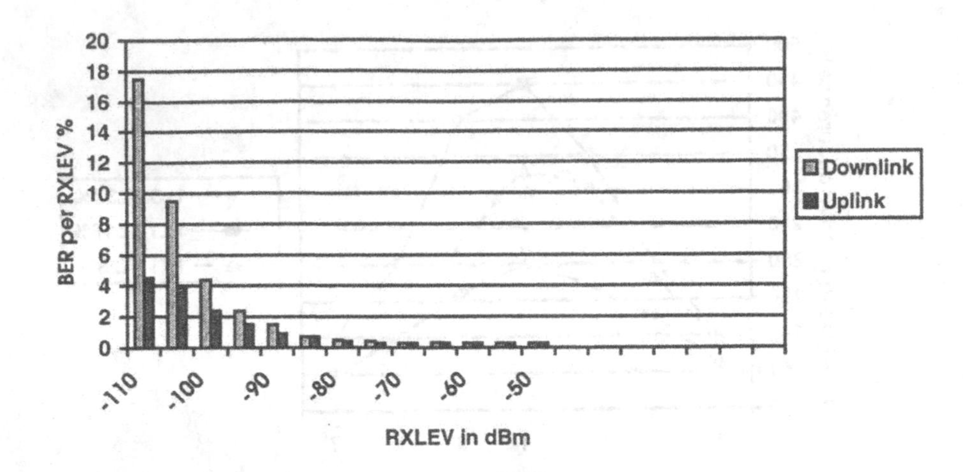

Figure 5. BER relationship with received signal level

The percentage of Audio mean samples for each RXQUAL level are shown in Figure 6 for the 1x1 SFH with 16% fractional load trials and compared with the case without Frequency Hopping. At RXQUAL levels up to 3 there is no perceptible difference between the hopped and non-hopped quality on the basis of the number of poor audio mean samples. The number of samples for RXQUAL 5 suggest that reasonably good audio quality is obtained at this level with frequency hopping but at level 6 the audio quality is indistinguishable for the hopped and non-hopped cases. This is useful in setting the lower RXQUAL threshold for power control i.e. power increase trigger level. The data from other scenarios also suggest that this behavior is reasonably consistent and that the threshold is not overly sensitive to the interference for the maximum fractional load.

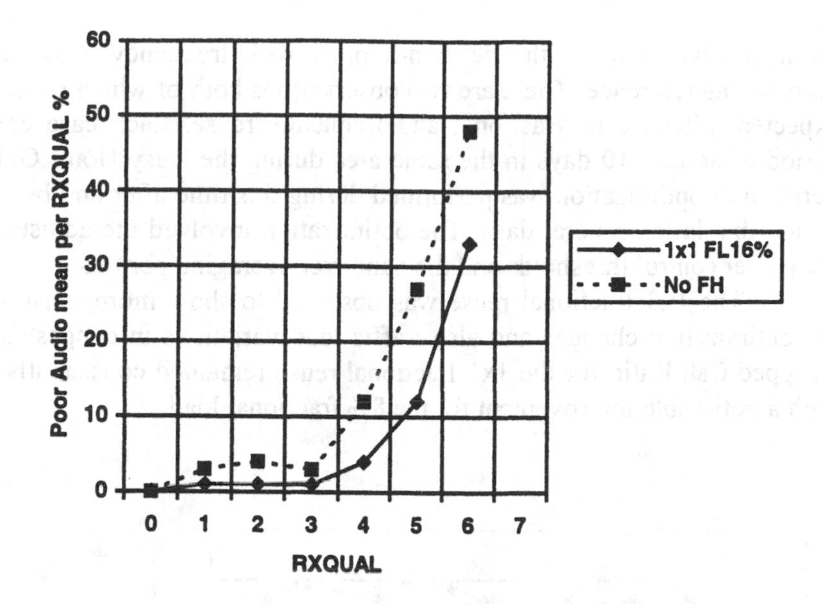

Figure 6. Voice Quality relationship with RXQUAL

System Performance statistics

The OMC counter data are routinely processed in all cellular systems to monitor system performance. Typically the following statistics are derived in most systems:

- Call Success Rate
- Handover Success Rate
- Handover Failure Rate
- Handover cause and attempts
- Dropped Call Ratio
- Traffic Volume
- Traffic and GoS

The detailed raw counters from which these statistics are derived can provide useful insight in the diagnosis of system malfunction or the occurrence of abnormal events. The Dropped Call Ratio and the Handover Attempts can indicate a change that alters the statistics for the Traffic Volume, Traffic and GoS. These statistics are considered for the fractional SFH systems for both trial and operational systems.

Dropped Call Ratio

The Dropped Call Ratio for the 1x3 and 1x1 fractional reuse are shown for different fractional load conditions in Figure 7. The results are presented on a relative scale with the non-hopped 4x3 frequency reuse as the normalising reference. There are two observations both of which confirm the expected influence of fractional and frequency reuse. Each case covers a period of at least 10 days in the same area during the Busy Hour. Only one iteration of optimization was performed during this time after an observation period that lasted several days. The optimization involved the adjustment of the power control thresholds and the handover averaging periods.

The 1x3 fractional reuse was observed to show more sensitivity to the optimization changes and also traffic load variations in congested cells. Dropped Call Ratio for the 1x1 fractional reuse remained consistently better with a noticeable improvement for the 8% fractional load.

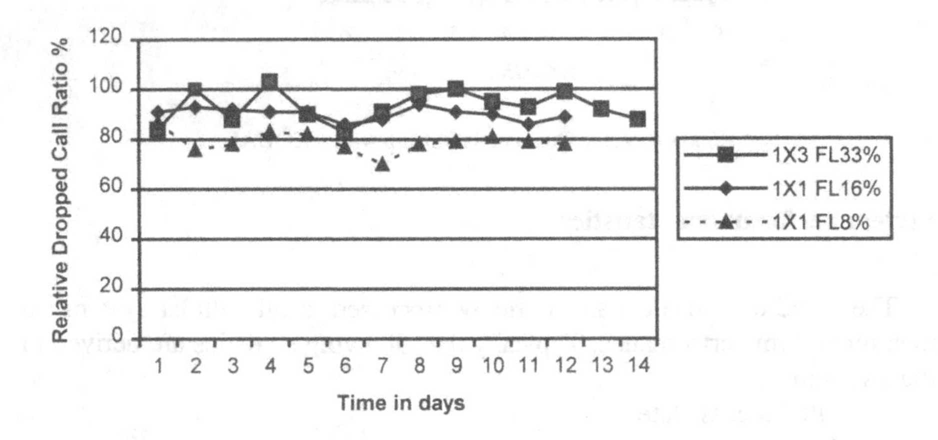

Figure 7. Dropped Call Ratio statistics for 1x1 fractional reuse

The 1x3 reuse reacted strongly to any changes in antenna orientation and to a lesser extent the vertical tilts. Changes in the neighbor cell topology, particularly with local congestion in some cells produced marked improvement in call quality. In this network the traffic directed handover was also activated and therefore the combined effect was to produce perceptible congestion relief.

The Dropped Call Ratio was observed to increase with increasing fractional load. In the 1x1 fractional reuse the statistics are consistently in favor of the lower load. These results should be treated with some caution, as the cells in this particular network were not in congestion.

Handover Attempts

The statistics for Handover Attempts for each handover cause are separately shown in Figure 8 for 1x1 fractional reuse with fractional load of 8 and 16%. The volume of handover attempts are noted to increase with frequency hopping and the proportion of handovers caused by poor quality is much higher compared to the non-hopping case. In typical non-hopping networks the thresholds for the handover parameters are set to trigger on power budget and typically these account for 80% of the handover causes. By some optimization of the cell parameters this can be re-balanced but the proportion of quality triggered handovers still remains larger than the non-hopped case. This is due to the changing characteristics of RXQUAL with frequency hopping.

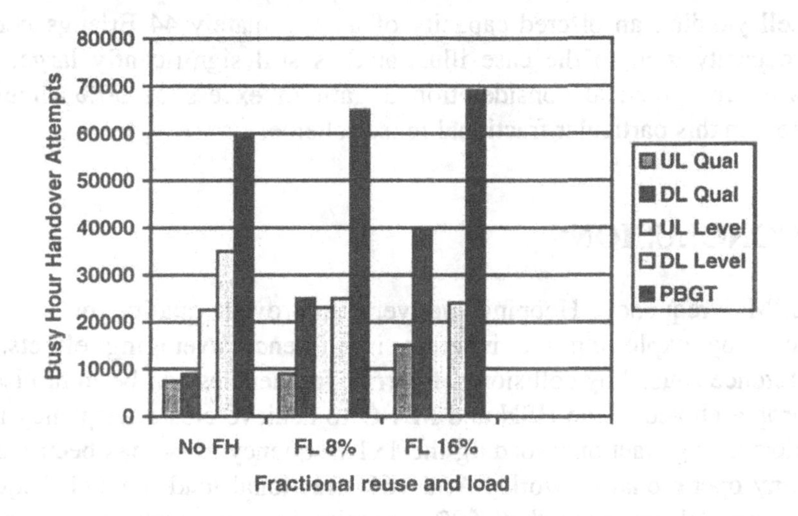

Figure 8. Handover Attempts statistics for 1x1 fractional reuse

The volume of handover attempts were reduced in successive iterations in the optimization by careful adjustments to the cell parameter thresholds, with detailed attention to the traffic and GoS. Poor optimization on the other hand can greatly increase 'ping-pong' effects with frequent and unnecessary handovers. This was observed in cases where the upper RXQUAL threshold was set too low causing premature handovers. The averaging period was also adjusted with favourable results in most cases. The 'ping-pong' effect can potentially cause increased dropped calls, especially where the congestion levels are high for a number of neighbor cells and the MAIO planning cannot guarantee sufficient interference margin.

Fractional SFH Capacity

The capacity with fractional reuse can be estimated using a simplified method. The practical results show that 1x1 fractional reuse is viable at a fractional load of 16%. The typical case in these trials of a spectrum allocation of 36 carriers, the number of TCH carriers available for frequency hopping is 24. Here we have assumed that 12 carriers are reserved for the 4x3 BCCH frequency reuse. At the 16% fractional load the number of hopping TRX per cell is 4 i.e. 24x0.16. Including the BCCH the total number of TRX equipped in each cell is 5. Allowing 4 SDCCH and BCCH control channels the number of TCH per cell is typically 36. At a GoS of 2% this equates to an offered traffic of 27.3 Erlangs per cell and 82 Erlangs per tri-sectored site. In comparison the 4x3 frequency reuse operates with 3TRX per cell yielding an offered capacity of approximately 44 Erlangs per site. The capacity gain in the case illustrated is still significantly large. Even allowing for practical consideration a gain in excess of 50% should be possible in this particular fractional reuse scheme.

4. CONCLUSIONS

GSM Frequency Hopping delivers improved quality or increased capacity by exploiting the inherent interference averaging effects. The interference caused by collisions of carrier frequencies can be minimised by the proper choice of the HSN and MAIO to achieve closer frequency reuse. By introducing fractional loading the 1x1 frequency reuse has been realized in many operational networks. At a 16% fractional load the 1x1 frequency reuse, can deliver more than 50% capacity increase compared to a non-hopped 4x3 frequency reuse and with improved overall voice quality. Data from trial systems and operational networks shows that the system performance can be maintained at the same time as increasing capacity.

The implementation of capacity enhancing features such as down link power control and DTX are generally beneficial in reducing interference. Frequency hopping in some cases can be combined with other traffic directed system features to improve the overall performance in traffic limited or interference limited networks.

System optimization with frequency hopping requires careful attention, as the thresholds for cell parameters need to be systematically adjusted to ensure good performance. In particular the thresholds that are triggered by RXQUAL are directly affected. Practical optimization experience indicates that reasonably consistent performance can be generally achieved. However, the optimization requires voice quality monitoring and FER analysis to ensure that consistent performance levels are maintained, especially as the traffic load increases and traffic re-balancing becomes necessary to maintain the frequency reuse over a wider area.

REFERENCES

[1] GSM Recommendations 05.02 and 05.03, ETSI.

[2] J.-L. Dornstetter and D. Verhulst, 'Cellular Efficiency with slow frequency hopping: Analysis of the digital SFH900 mobile system', IEEE J. Select. Areas Comm.., vol.SAC-5, no. 5, pp.835-848, July 1987.

[3] S. Chennakeshu, et al., 'Capacity Analysis of TDMA-Based Slow-Frequency-Hopped Cellular System', IEEE Trans.Veh. Technol., vol.45, no. 3, pp.531-542, Aug. 1996.

[4] B. Gudmundson, J. Skold, and J.K. Ugland, 'A comparison of CDMA and TDMA sysytems', in Proc. IEEE Veh. Tech. Conf., Denver, CO, May 1992, pp.400-404.

PART IV

DEPLOYMENT OF WIRELESS DATA NETWORKS

Chapter 10

GENERAL PACKET RADIO SERVICE (GPRS)

Fixed Deployment Considerations

DR. HAKAN INANOGLU**, JOHN REECE*, DR. MURAT BILGIC*

**Omnipoint Technologies Inc. **Opuswave Networks Inc.*

Abstract: The General Packet Radio Service (GPRS) is the first evolutionary step, in deploying a truly mobile wireless internet capability, for GSM and TDMA operators. As an upgrade to currently deployed networks, operators providing GPRS must be able to provide this service, with acceptable quality, within the physical constraints of the existing system infrastructure. As a result, it is imperative that the operator's technical and marketing personnel be cognizant of the difference in performance characteristics, at the physical layer, between GPRS and GSM or TDMA. This chapter identifies the physical layer characteristics, and expected system performance, for slow moving and stationary terminal units.

1. INTRODUCTION

GPRS

General Packet Radio Service (GPRS) is an overlay extension for the GSM network to provide packet-based communication. It is designed to carry Internet Protocol (IP) and X.25 traffic destined to/from Terminal Equipment (TE) accessing the Wide Area Network (WAN) through a GSM wireless connection.

Services

There are two categories of GPRS services as defined in [1], Point To Point (PTP) services and Point To Multipoint (PTM) services.

PTP services have two flavors, a connectionless network Service (PTP-CLNS) to carry IP traffic, and a connection oriented network service (PTP-CONS) to carry X.25 traffic.

As described later, GPRS is being introduced in phases. In the first phase, the focus is on PTP services. In the second phase, Point-to-Point Protocol (PPP) shall be added as a separate Packet Data Protocol (PDP) type to be carried over GPRS. In addition to PTP services, GPRS provides Short Message Service (SMS) transfer over GPRS radio channels.

Architecture

GPRS is based on the use of new GPRS radio channels. The allocation of these channels is flexible such that from 1 to 8 radio timeslots can be allocated independently, for uplink and downlink per TDMA frame period. The radio interface resources can be shared dynamically between existing circuit-switched services and GPRS services.

Cell selection may be performed autonomously by a Mobile Station (MS), or the Base Station System (BSS) instructs the MS to select a certain cell. The MS informs the network when it re-selects another cell or group of cells known as a routing area.

GPRS introduces two new network nodes in the GSM Public Lands Mobile Network (PLMN). The Serving GPRS Support Node (SGSN) keeps track of the location of each MS in its routing area and performs security functions and access control. The SGSN is connected to the BSS typically via Frame Relay. The Gateway GPRS Support Node (GGSN) provides interworking with external packet-switched networks, and is connected with SGSN(s) via an IP-based GPRS backbone network. The Home Location

Register (HLR) is enhanced with GPRS subscriber information, and the SMS nodes are upgraded to support SMS transmission via the SGSN. Reference [2] describes GPRS logical architecture in detail.

GPRS security functionality is equivalent to the existing GSM security. The SGSN performs authentication and cipher setting procedures based on the same algorithms, keys, and criteria as in existing GSM. GPRS uses a new A5 ciphering algorithm optimized for packet data transmission.

To access GPRS services, a MS first makes its presence known to the network by performing a GPRS attach. This operation establishes a logical link between the MS and the SGSN, and makes the MS available for SMS over GPRS, paging, and notification of incoming GPRS data

To send and receive GPRS data, the MS activates the packet data address that it wants to use. This operation makes the MS known in the corresponding GGSN, and interworking with external data networks can commence.

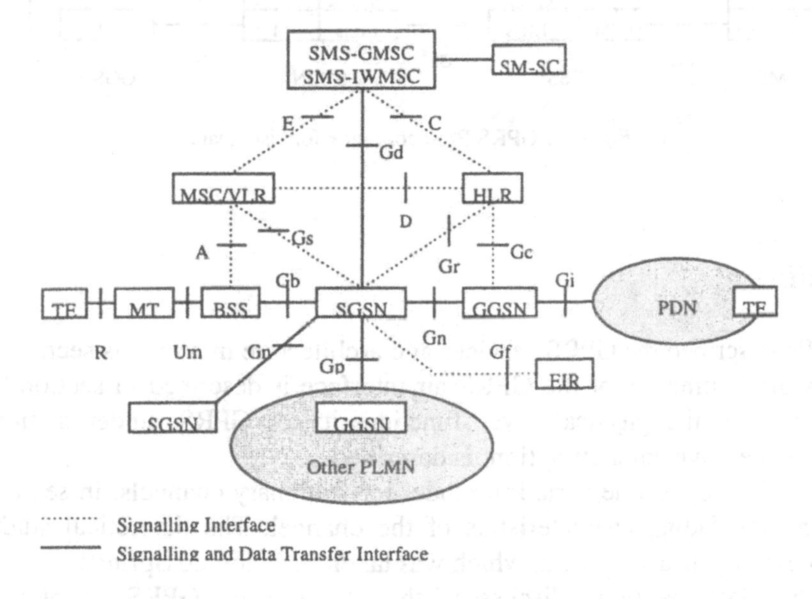

Figure 1. Overview of GPRS Logical Architecture

GPRS Tunnelling Protocol (GTP) is used to tunnel both IP and X.25 traffic. GTP can be carried over Transmission Control Protocol (TCP) for X.25 traffic and over User Datagram Protocol (UDP) for Internet Protocol (IP) traffic. User data packets are encapsulated in Sub-Network Dependent Convergence Protocol (SNDCP) Protocol Data Units (PDUs) which are carried over Logical Link Control (LLC) layer. The LLC is designed to be radio network independent so that GPRS can be used over different radio networks. LLC has acknowledged and unacknowledged modes. BSS GPRS Protocol (BSSGP) carries routing and QoS information between the SGSN and the BSS. The Radio Link Control (RLC) provides a radio-solution-dependent reliable link, whereas Medium Access Control (MAC) controls the access signalling procedures for the radio channel, and the mapping of LLC frames onto the GSM physical channel. The RLC also provides both acknowledged and unacknowledged modes of transmission.

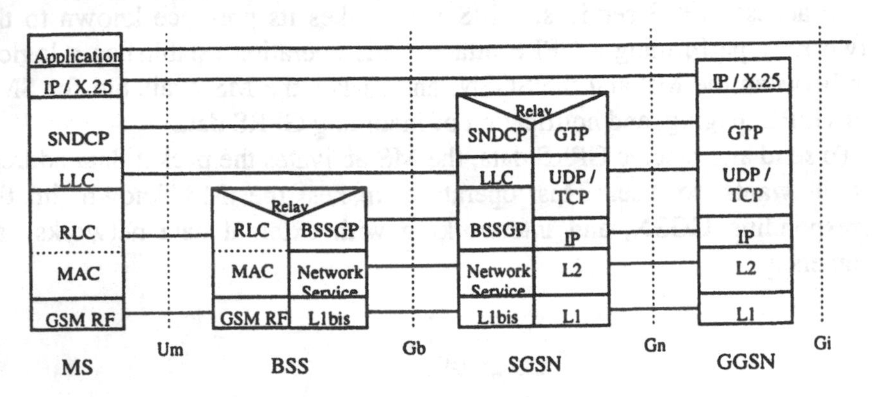

Figure 2. GPRS Protocol Stack for User Data

Outline

We described the GPRS services and architecture in previous section.

A brief summary of the GPRS air interface is described in section 2. In this section, the physical layer functionality of GPRS, under stationary channel deployment assumption, is described.

We develop a time variation model for stationary channels, in section 3, to find the fading characteristics of the channel. The theoretical study is supported by measured data, which was taken in Colorado Springs.

The last section discusses the traditional GPRS deployment considerations. First, we introduce the coexistence of GPRS with other cellular systems in the US Personal Communication System (PCS) band, and then we define various interference sources that can degrade radio system performance and reduce availability, coverage and capacity. Next, we summarize the classical GSM deployment that is applicable to fixed GPRS deployment. Finally, we analyze the impact of co-channel and adjacent channel interference on cell availability, for both indoor and outdoor terminals. The analyses are performed for quasi stationary and stationary cases with and without shadow fading. The results are provided for different GPRS channel codes.

Finally, we conclude our investigations by addressing GPRS fixed channel deployment issues.

2. GPRS AIR INTERFACE

The GPRS air interface is an overlay to the existing GSM air interface. This is accomplished by introducing new GPRS logical channels. Therefore, to describe GPRS air interface characteristics, we first need to introduce the new logical channels. Then we will describe the radio resource management, i.e., the GPRS channel allocation in a given cell, as well as the modes of the mobile. Then we discuss the GPRS specifics of the physical layer, such as coding, cell reselection, timing advance, and power control.

Physical Layer

Channel Coding

Four coding schemes, CS-1 to CS-4, are defined for PDTCH. For all packet control channels, except PRACH and uplink PTCCH, CS-1 is used. For PRACH and PTCCH uplink, the coding scheme used for GSM random access bursts is used.

The coding procedure starts by adding a Block Check Sequence (BCS) for error detection. Then the following steps are taken for each coding scheme:

a) CS-1: Utilises a half-rate convolutional coder. The raw data rate of CS-1 is 9.05 Kbps.
b) CS-2: The effective coding rate is close to 2/3. The raw data rate of CS-2 is 13.4 Kbps.
c) CS-3: The effective coding rate is close to 3/4. The raw data rate of CS-3 is 15.6 Kbps.
d) CS-4: There is no Forward Error Correction (FEC) for CS-4. The raw data rate of CS-4 is 21.4 Kbps.

Cell Reselection

Cell reselection is equivalent to handover for circuit-switched services. However, unlike handover, the MS performs the cell reselection. New cell reselection criteria are defined in addition to existing GSM cell reselection criteria.

In GPRS, there is a provision for the network to request measurement reports from the MS. In this case, the network performs the cell reselection.

In stationary environments, cell reselection is a less likely event compared to the environments where the MS is highly mobile. This reduces the need for frequent measurements on the target cell BCCH/PBCCH frequencies. Therefore, it's possible to schedule longer bursts of transmission when a multi-slot MS, e.g., 8-slot, is used. This reduces the overall delay for the transmission of long IP packets.

Timing Advance

The MS uses the timing advance procedure to obtain the timing advance value for uplink transmissions. When the MS sends an access burst carrying the Packet Channel Request, the network makes an initial timing advance estimation. This estimation is sent back to the MS in a Packet Uplink or Packet Downlink Assignment.

In stationary environments, the timing advance can be calculated and stored in the MS. Therefore, the initial timing advance estimation should be fairly accurate in such environments. This will help start the data transfer as early as possible.

Power Control

For the uplink, the MS uses a power control algorithm, which can be in both open loop and closed loop modes. The output power is calculated as a function of a frequency band constant (Γ_0), and channel specific power control parameter sent in the resource assignment message (Γ_{CH}), and the received signal level at the MS. The MS output power is limited by the maximum allowed output power in the cell (PMAX). When accessing the PRACH, the MS always uses PMAX.

As we will show in the next section, the maximum fade depth in a static channel is approximately 8 dB. Therefore, power control threshold parameters should be set such that the final power level will be at least 8 dB above the receiver sensitivity level. The slow varying characteristic of stationary channels will yield very good performance for the GPRS power control technique.

3. STATIC CHANNEL TIME VARIATION

The propagation channel characteristic has a great impact on overall radio system performance. Capacity, interference, range and Quality of Service (QoS) parameters vary significantly from benign to severe channels. In this section, we will explain the time variation characteristics of the static propagation channel. Here the static channel means that the transmitter and receiver locations don't move. Therefore, moving obstacles, in between the transmitter and receiver locations, cause the received signal time variation. In this section, we will show that the time variation of a static channel is very slow, and the maximum fade depth is much less than the one that can be seen in fast fading channels with high mobility.

Mathematical Modeling

The moving objects, such as cars and people, between transmitter and receiver, are the only reason for time variation in the static channel. As the majority of the movements take place at the street level, they modulate the phase and amplitude of the ground reflected rays. The phase and amplitude of the other rays, reflected or diffracted by other objects above ground level, will not vary with time, as far as the reflection points don't move.

The static channel time variation will be analyzed for a worst case scenario where there is no line of sight radio link between the transmitter and receiver locations. The physical model used, in derivation of the equation, to investigate the time variation of the static channel, is depicted in *Figure 3*. The figure shows a terminal receiver, installed on the outer wall of the second building, at height of h_t measured from the ground. The time variation of the received signal level will be investigated when a car passes between the two buildings. As there is no line of sight component of the rays coming to the receiver, and the ground reflected ray is obstructed by a high conductance obstacle the top surface of the car the scenario shown in *Figure 3* reflects the worst case time varying scenario.

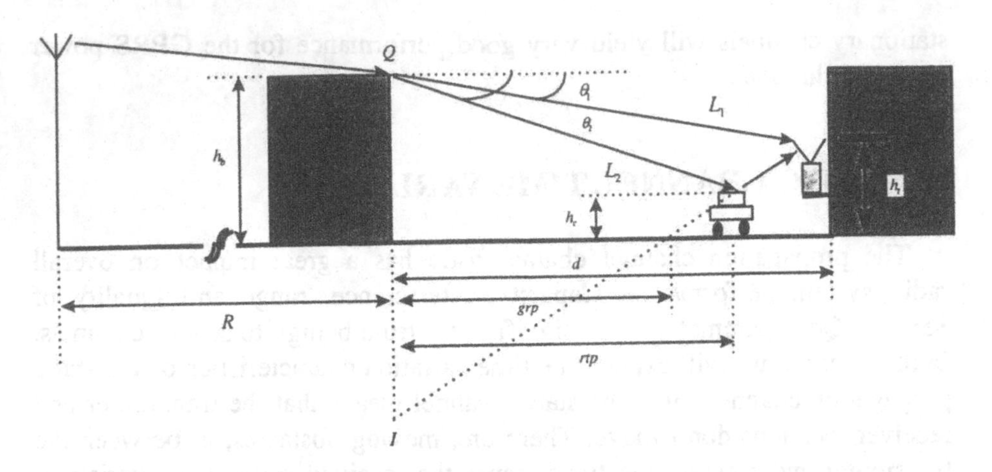

Figure 3. Stationary Channel Time Variation Model

In *Figure 3*, the distance between the transmit antenna and roof top diffraction point, Q, is represented by R. The height of the first building, and the car measured from the ground, are h_b and h_c, respectively. There are two rays coming to the terminal location after the incoming field is diffracted from diffraction point, Q. As is depicted in the figure, the first ray path, L_1, reaches the terminal without additional reflection or diffraction. However, the second ray, L_2, is reflected from the ground, in the case where no car is passing between the two buildings. The distance of the diffraction point, to the ground reflection point, is grp. A car passing very close to the terminal location will obstruct the ground-reflected ray. However, the second ray will be reflected from the conductive car rooftop changing the reflection distance to rtp meters. The diffraction point will behave as a line source generating cylindrical waves down to the street level. The reflection distance, with and without the car, can be calculated by use of the image location of the line source, I. If the total distance traveled by the first ray is r_1 and the diffraction angle of the first ray is θ_1 then the path loss of the first ray component will be as follows

$$L_1 = \frac{1}{2\pi k r_1}\left[\frac{1}{|\theta_1|} - \frac{1}{2\pi + |\theta_1|}\right]^2 \tag{1}$$

Here, k is the propagation constant and r_1 is the total distance traveled by the first and second rays from the diffraction point to the terminal. The distance can be calculated in terms of the given geometry as follows

$$r_1 = \sqrt{d^2 + (h_b - h_t)^2} \tag{2}$$

The diffraction angles θ_1 is

$$\theta_1 = \tan^{-1}\left\{\frac{h_b - h_t}{d}\right\} \tag{3}$$

The second diffraction ray, with angle θ_2, gives the reflected path excess path loss given by

$$L_2 = \frac{\Gamma^2}{2\pi k r_2}\left[\frac{1}{|\theta_2|} - \frac{1}{2\pi + |\theta_2|}\right]^2 \tag{4}$$

Here Γ is the reflection coefficient of the car rooftop surface. The total distance from the diffraction point to the terminal (r_2) is given by the following equation

$$r_2 = \sqrt{d^2 + (h_b + h_t)^2} \tag{5}$$

The diffraction angle (θ_2) for the second ray is

$$\theta_2 = \tan^{-1}\left\{\frac{h_b + h_c}{rtp}\right\} \tag{6}$$

Representing the ratio of the received power levels of the two rays with Δ, one can calculate the maximum fade depth in dB as

$$F_{max} = 10Log_{10}\left\{\frac{1}{1-\Delta}\right\} \qquad (7)$$

The variation of the power level ratio (Δ) and the maximum fade depth is given in *Figure 4* as a function of the receiver terminal height.

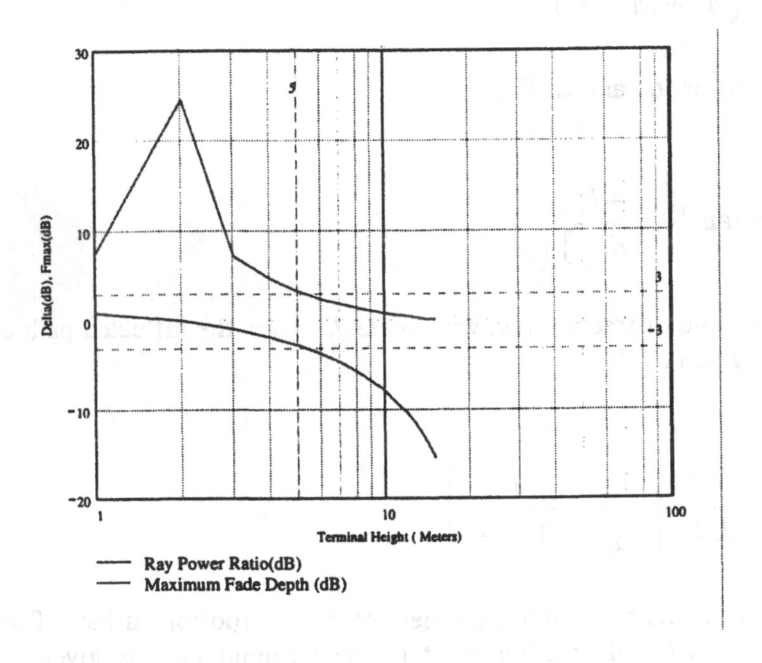

Figure 4. Terminal Height versus Fade Depth Variation of a Stationary Channel

The time variation of the received signal level measured at the intersection of Nevada / Platte Avenue, in Colorado Springs / Colorado, is depicted in *Figure 5*. In this measurement, the transmitter antenna height was 57 meters, and receiver antenna height was 3 meters. The maximum fade depth within a 60 second test duration, was measured as 8 dB, yielding perfect agreement with the analysis results.

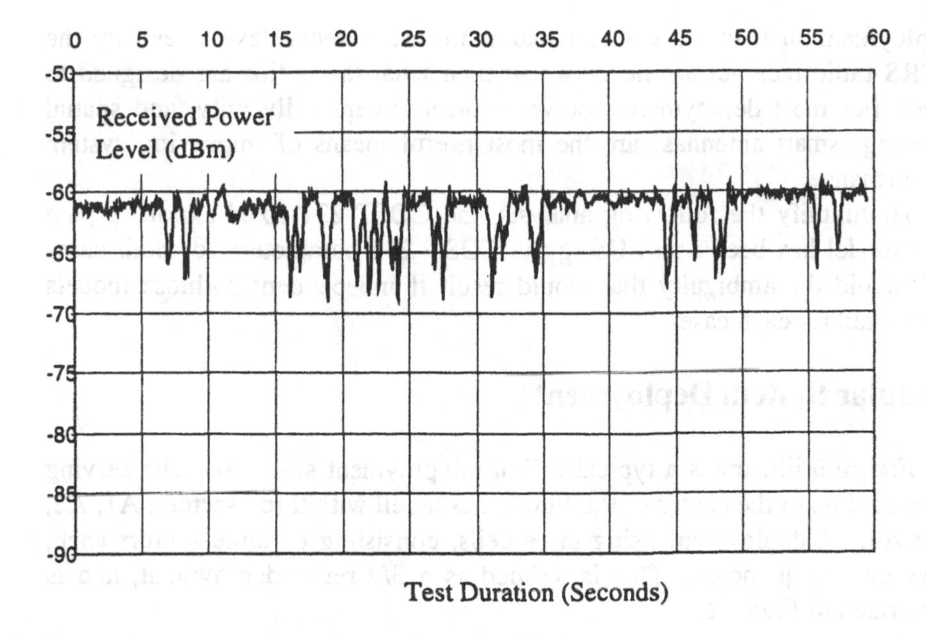

Figure 5. Colorado Springs Static Channel Time Variation Measurement

4. GPRS DEPLOYMENT CONSIDERATIONS

Introduction

At the radio interface, GPRS must co-exist in a radio environment that is polluted by other systems using the same or nearby radio frequency resources. In particular, for the United States PCS band, these other radio resources include both co-located and non co-located GPRS/GSM, TDMA (IS-54), and CDMA (IS-95) base stations and mobiles. Interference from these sources can result in performance degradation of the GPRS radio receiver through such mechanisms as co-channel wideband phase noise and modulated carrier power, reciprocal mixing, in-band intermodulation and spurious products, and high level blocking of the receiver front end.

The GPRS system requirements specify the radio receiver performance, given a maximum input interference power level, for co-channel, adjacent channel, and intermodulation, and spurious products. This document is used for the design of the radio receiver, but it is a practical matter to design the deployment such that the actual maximum interference levels seen by the GPRS radio receiver are never worse than what the radios are designed to meet. For most deployments, power control, antenna diversity, and spatial filtering (smart antennas) are the most useful means of improving system performance.

To simplify the following analysis, the COST 231 fixed exponent path loss model has been used. Using the COST 231 parameters for both cases will avoid the ambiguity that would result if independent pathloss models were used for each case.

Cellular System Deployment

Figure 6 illustrates a typical cellular deployment scenario. The serving base station, in the center of the figure, has a cell with three sectors, A1, A2, and A3. A deployment using three cells, consisting of three sectors each, uses nine frequencies. This is defined as a 3/9 reuse deployment, and is illustrated in *Figure 6*.

The mean co-channel C/I ratio, for a terminal unit within the serving cell (A) in *Figure 6*, is proportional to the ratio of the distance d_s from the serving base, to the distance d_i from the interfering base, such that

$$\frac{C}{I} = \left[\frac{d_s}{d_i}\right]^{-\alpha} \tag{8}$$

Where α is the path loss coefficient. Placing the terminal unit at the edge of the cell results in $d_s = 2R$ and $d_i = D = 6R$. This results in

$$\frac{I}{C} = \left(\frac{D}{R} - 1\right)^{-\alpha} \tag{9}$$

It can be shown, for a hexagonal arrangement of cells, the factor $D/R = \sqrt{3N}$, where N is the number of frequencies used in a cluster. Thus the log value of the carrier-to-interference ratio can be cast in the form,

$$C - I = 10Log_{10}\left(\sqrt{3N} - 1\right) \tag{10}$$

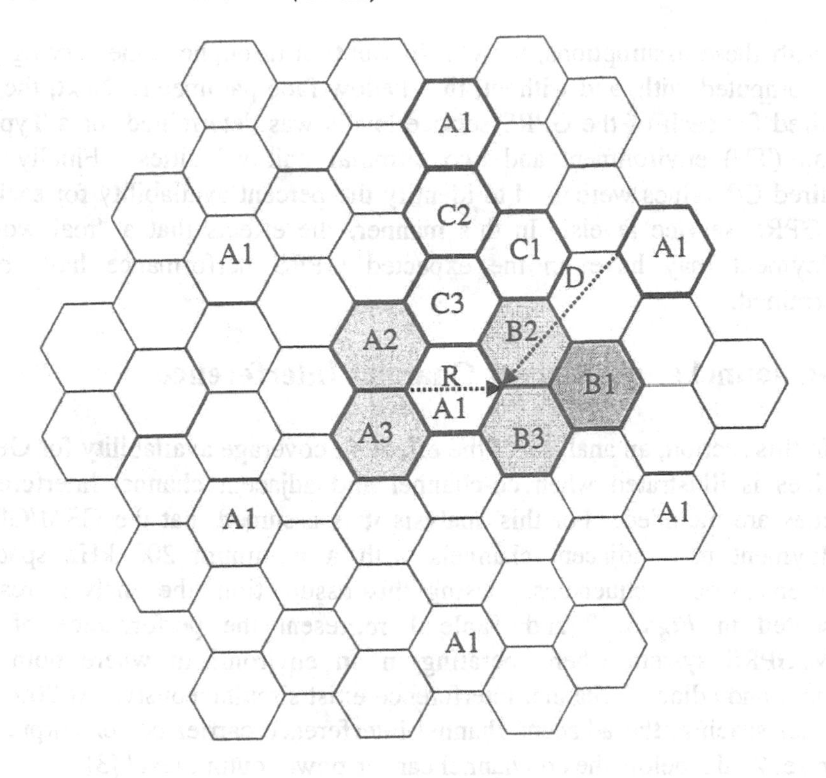

Figure 6. Typical Cellular Hexagonal Deployment Geometry

Using a 3/9 deployment, and substituting $\alpha = 3.5$ (from the COST 231 Modified Hata Model at 1900 MHz) into equation 10, results in a minimum C/I of 10.5. For a 4/12 deployment, the minimum expected C/I is 13.7 dB.

Since spectrum is a finite resource, and reducing GSM system capacity would impact customer satisfaction, many operators will have to re–optimize their radio networks for a minimal frequency reuse scheme. For some operators, this could mean reducing the reuse from 7/21 to 3/9. The resulting decrease in the minimum expected C/I, under these circumstances, is approximately 9 dB. Since the co-channel C/I required for acceptable GSM frame erasure rates, in standard GSM voice operation, is 9 dB, the C/I determined for a 3/9 reuse scheme is minimally sufficient, without margin, to provide acceptable service at the cell edge.

For ease of computation, the analysis carried out here assumes that all signals are totally decorrelated from each other and that the log normal standard deviation of the shadow fade, for the interference sources, is one-half of the standard deviation of the serving cell, for the outdoor terminal unit.

With these assumptions, the C/I distribution throughout the serving cell was computed with, and without, the shadow fade parameter. Next, the C/I required for each of the GPRS service levels was determined for a Typical Urban (TU) environment and two terminal unit velocities. Finally, the required C/I values were used to identify the percent availability for each of the GPRS service levels. In this manner, the effects that a "real world" deployment may have on the expected GPRS performance have been determined.

Co-Channel and Adjacent Channel Interference

In this section, an analysis of the effect on coverage availability for GPRS services is illustrated when co-channel and adjacent channel interference sources are included. For this analysis it is assumed that the GSM/GPRS deployment uses adjacent channels with a minimum 200 kHz spacing between carrier frequencies. Using this assumption, the analysis results presented in *Figure 7* and Table 1 represent the performance of the GSM/GPRS system when operating in an environment where both co-channel and adjacent channel interference exist simultaneously. At 200 kHz channel spacing, the adjacent channel interference carrier power output has been set 27 dB below the co-channel carrier power output level [3].

Table 1 shows that the availability of the GPRS CS-4 service level is 44% for the 1.5 km/hr channel and 29% for the 50 km/hr channel when the shadow fade parameter is included. These results show that even though the C/I requirement for GSM voice service results in an acceptable 93% availability, the high data rate GPRS service suffers from the need for a higher overall C/I throughout the coverage area.

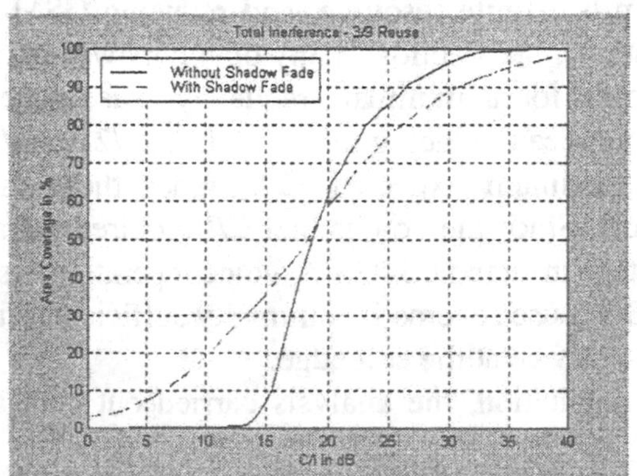

Figure 7. GSM/GPRS Cell Availability versus C/I for Co-Channel and Adjacent Channel Interference

Table 1. GPRS Availability with and without Shadow Fading Parameter for Co-Channel and Adjacent Channel Interference

GPRS Service Level	Required C/I [dB]	Area of Cell Availability (%) No Shadow Fade	Area of Cell Availability(%) With Shadow Fade	
	TU-1.5/TU-50	TU-1.5/TU-50	TU-1.5 Channel Model	TU-50 Channel Model
CS-1	13.0/9.0	99/99	70	83
CS-2	15.0/13.0	92/99	64	70
CS-3	16.0/15.0	77/92	62	64
CS-4	19.0/23.0	42/20	44	29

Co-Channel and Adjacent Channel Interference for Indoor Mobile

Because of reciprocity, the results presented in the last section are applicable for both the downlink and uplink communication paths, when the terminal unit is outdoors. For a terminal unit indoors, the mean received carrier and interference power levels are reduced by the amount of penetration loss into the building. Since the pathloss to a terminal unit within a building can vary greatly, a larger lognormal standard deviation is used to model the expected pathloss for this case. To accurately model the indoor terminal case, the indoor downlink should be modeled with a slow or static channel model.

The Block Error Rate (BLER) versus C/I for an indoor terminal unit, in a static channel, have been derived. Using the average difference of the required Eb/No and C/I, at a BLER of 10%, in the TU-3 and TU-50 channel models. The average value of the difference calculated is approximately 1.5 dB. Therefore, by adding 2 dB to the published static channel Eb/No requirement, at 10% BLER, representative values of the required C/I for each GPRS service level can be calculated.

The calculated C/I distribution, for the downlink path, to an indoor GPRS terminal was calculated as in the last section. The results were used to calculate the GPRS availability, on the downlink path, to an indoor terminal unit. These results are presented in Table 2. For the shadow faded downlink, the availability of the GPRS CS-4 service level is 58% in a static channel, and only 35% for the slow pedestrian (1.5 km/hr) channel.

Table 2. The Downlink GPRS Cell Coverage Area for an Indoor Terminal Unit, with and without Shadow Fading Parameter, for Co-Channel and Adjacent Channel Interference

GPRS Service Level	Required C/I [dB]	Area of Cell Availability (%) No Shadow Fade	Area of Cell Availability(%) With Shadow Fade	
	Static/TU-1.5	Static/TU-1.5	Static Channel Model	TU-1.5 Channel Model
CS-1	3.0/13.0	100/99	72	48
CS-2	6.0/15.0	100/94	67	43
CS-3	7.0/16.0	99/90	64	42
CS-4	10.0/19.0	99/52	58	35

For the indoor terminal unit uplink path, the desired signal is treated the same as for the downlink, but the interference terminals cannot be constrained to an indoor, static location. From a pathloss standpoint, the worst case condition, is when the desired unit is indoors and the interfering units are all outdoors. The resulting C/I distribution is biased to lower values because the carrier power is reduced, by the penetration loss, while the interference power levels are not. The effect of this reduction in C/I is evident in the availability of each GPRS service level, as shown in Table 3. For this scenario, the availability for the lowest rate GPRS service (CS-1) is 53% for the static channel and 28% for the 1.5 km/hr pedestrian channel.

Table 3. The Up-link GPRS Cell Coverage Area for an Indoor Terminal Unit, with and without Shadow Fading Parameter, for Co-Channel and Adjacent Channel Interference

GPRS Service Level	Required C/I [dB]	Area of Cell Availability (%) No Shadow Fade	Area of Cell Availability(%) With Shadow Fade	
	Static/TU-1.5	Static/TU-1.5	Static Channel Model	TU-1.5 Channel Model
CS-1	3.0/13.0	68/9	53	28
CS-2	6.0/15.0	45/5	48	23
CS-3	7.0/16.0	35/4	44	22
CS-4	10.0/19.0	18/1	35	15

5. CONCLUSIONS

In this study, we introduced the GPRS network architecture and air interface features. We proved that, even in a worst case scenario, the time variation of stationary channel is very slow and the fade depth is far less than the fade depth of a mobility system. Therefore, the stationary channel can be approximated by a Gaussian channel in most cases. Although the physical layer functions such as Timing Advance, Cell Re-Selection and Power Control are designed mainly for a mobility system, we have shown that these features will show improved performance for fixed deployments.

It has been shown that the worse case links for GPRS are for indoor terminal units. The analysis shows that for an indoor terminal unit, at pedestrian velocity, the downlink availability is 42% to 48% for the CS-1 through CS-3 service levels, and 64% to 72% for the static case. The indoor terminal uplink availability for the CS-1 through CS-3 service levels is 22% to 28% for the slow pedestrian velocity and 44% to 58% for the static case. For the indoor terminal an improvement in the link availability of 150% is achieved for the downlink static versus pedestrian velocity results. Similarly, for the uplink availability, an increase of nearly 200% is seen for the static versus pedestrian velocity results.

REFERENCES

[1] GSM 0260: "GPRS Service Description, Stage 1" Ver 7.1.0, April 1999
[2] GSM 03.60: "GPRS Service Description, Stage2"Ver 7.0.0, April 1999
[3] GSM 05.05: "Radio Transmission and Reception" Ver 6.2.0, 1997

Chapter 11

WIRELESS LAN DEPLOYMENTS: AN OVERVIEW

CRAIG J. MATHIAS

Farpoint Group

Abstract: Wireless LANs allow LAN services to be delivered without the need for a wired connection between the client and the supporting infrastructure. Most products today use some form of spread-spectrum microwave transmission, typically operating in unlicensed bandwidth. In addition, most products are based on a microcellular infrastructure, allowing roaming through arbitrarily-large areas. With the emergence of the extensible IEEE 802.11 standard, the number of available products has increased dramatically, and products based on second-generation 802.11 PHYs are offering throughput commensurate with wired LANs. Issues related to deployments are normally limited to the placement of access points according to the coverage desired and the constraints of specific in-building RF propagation, educating users with respect to antenna location and orientation, and interference management.

1. WHAT IS A WIRELESS LAN?

A wireless LAN (WLAN) is just that - a LAN implemented, completely, or for the most part, without wires. In most cases, the wireless component will consist of the connection between an *end node* or *station* in a given network (e.g., a PC, typically mobile) and a bridge (usually called an *access point*, or *AP*) between the wireless connection and a wired infrastructure. Other than being implemented without a wired physical connection, all typical LAN functionality is preserved. Indeed, only the physical layer with respect to the ISO-OSI open systems model of networking need be affected, although, in practice, most implementations require a data link layer (via appropriate driver software) specific to a given implementation of wireless LAN and a particular network operating system (NOS). All higher-level functionality (provided by the NOS and appropriate protocol stacks) is maintained. No other changes to the software configurations of network nodes are normally required. In this sense, wireless LANs are a "plug and play" replacement for wired LANs.

Justifications for Wireless LANs

Just when wireless LANs can (and should) be substituted for their wired counterparts are a source of much confusion. In general, wireless LAN components cost more than their wired LAN counterparts; this is due to both the greater level of technology required and the lack of sufficient product volumes so as to reduce component costs as a result of mass production. In fact, the claim that wireless LANs are simply too expensive to be applicable in many cases has been a major barrier to their adoption. While wired LAN network interface cards (NICs) typically cost between $20 and $100, their wireless counterparts typically have retail prices in the $100 to $300 range, plus access points which typically cost from $800 to $1,200 each. Regardless, wireless LANs can be significantly less expensive to maintain than their wired equivalents, primarily due to the reduced costs with respect to the moves, adds, and changes which are a core component of the life cycle costs of almost every LAN installation. Thus it is possible to recommend wireless LANs is many installations, even those not involving mobility, strictly on the basis of life-cycle cost.

Installation-Limited Situations

Beyond simple cost justification, the second major application for WLANs is in situations where wire cannot be easily (or otherwise) installed at all. Even in structures with ample room for cabling, situations often arise where it is just too difficult to install cable in certain locations. Some example of these include:

- *Factory and shop floor*, where cabling could interfere with other equipment, or where RF noise could affect baseband cabled networks.
- *Remote sites and branch offices*, where on-site networking expertise might not be available. Computers equipped with wireless LANs can be pre-configured and shipped ready to use.
- *Retail stores*, where modern communicating cash registers are required for stock management and real-time financial reporting, Most retail settings are designed for reconfigurability, with plenty of AC outlets located in concrete-slab floors, but most older sites have no data cabling.
- *High-security applications*, where precautions against cables being tapped (either physically of via RF emissions) must be taken. In this case, wireless LAN systems based on infrared (IR) are often used, as IR does not (unlike RF and microwaves) penetrate solid objects like walls.
- *Physical obstructions*, such as metal or reinforced-concrete walls, or to bridge across atria or ornamental spaces.
- *Environmental hazards*, such as lead paint or asbestos. These materials are often assumed to be safe if not disturbed, as would be the case with the installation of cabling.
- *Historic buildings*, which may have marble floors, paneled walls, or other construction-related issues.
- *Disaster recovery*, where pre-configured equipment can be rapidly deployed, or as a hot- or cold-standby measure in mission-critical settings.

Finally, it may indeed be the case that there is simply no more room for any new cabling even in buildings, which might once have had ample room for cabling. In this situation, wireless LANs are the LAN of last resort. It should be noted that despite the broad range of venues appropriate for WLANs, it is generally recommended that wired connections be used unless a WLAN system can be clearly justified on the basis of life-cycle cost, physical restrictions, or, as we shall discuss shortly, the need for mobility.

Vertical and Horizontal Applications

Perhaps due to the relatively greater cost of wireless LAN hardware, and their historically lower performance, the majority of installations to date have involved applications in vertical markets, where costs are often easier to justify. These include healthcare, retail, warehousing, distribution, manufacturing, education, and financial services

Most wireless-LAN vendors (and many market researchers and industry analysts) are assuming the emergence of a broad horizontal market, addressing typical LAN applications in office (and, increasingly, *residential*) settings, such as file and peripheral sharing, shared data-base and applications access, and access to the Internet and corporate intranets. The development of this market has been held back in recent years by the relative expense of wireless LANs, the lack of industry standards, and poor marketing on the part of the wireless-LAN industry. With all three of these issues now addressed, wireless LANs could become the norm in many settings.

In summary, wireless LANs are best utilized where wire cannot be *economically justified*, *physically applied*, or where the *dynamic mobility* of end nodes is a requirement. If one or more of these conditions are not satisfied, users are usually better off with a wired LAN. One advantage that wire seems destined to maintain over wireless is in throughput, although the gap is narrowing, and perceived performance (on the part of users) remains more important than benchmark results which can vary widely in wireless networks due to continual variations in the radio environment.

Microcellular Networks

Most significantly, wireless LANs open up the possibility of users remaining connected to a LAN infrastructure while dynamically roaming through a department, building, or even a campus setting and beyond. The rapid rise of mobile computers as the information processor of choice for many professionals was driven largely by the ability of the mobile computer to function in the office, home, or while travelling - in short, from a computational perspective, the user is location-independent. However, with the utility of the mobile computer largely defined by the information provided to it via a LAN, a mobile computer disconnected from the network, no matter how powerful, is of significantly reduced value. As a consequence, otherwise mobile computers remain tethered to office walls, despite the fact that their users may be away from their desks much of the day.

Mobility-oriented wireless LANs are typically implemented via a combination of a fixed infrastructure [figure 1] implemented with *access*

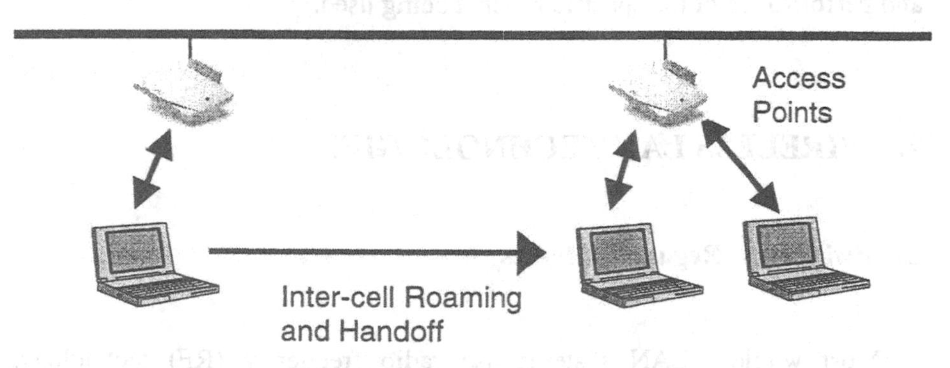

Figure 1 - Microcellular Roaming

points distributed around a building or even campus or courtyard setting, and a mobile wireless-LAN adapter for each roaming PC, typically implemented as a PC Card (formerly known as a PCMCIA card). Most PC Card implementations are in the Type II form factor, and increasingly have a one-piece design, with an integrated end-cap antenna. Some models allow for a variety of antenna configurations, and some use a two-piece design, with the RF electronics or dipole antenna separated from the PC Card and attached via an adhesive to the top surface of the notebook computer.

Wireless LANs designed for mobility are often referred to as having a *microcellular* architecture. In this case, these systems behave very much like a cellular telephone network, handing off moving users between and among access points as a user roams out of range of one access point and into the coverage area of another. Placement of access points can be a demanding exercise in trial-and-error, given the vagaries of RF propagation, especially indoors, and the frequent need to provide essentially uniform coverage of what might be a relatively large area. Most products include at least a rudimentary form of "site-survey" tool, which allows an individual to roam through a given area measuring the quality of an RF signal from an access point, but more sophisticated tools are under development. For example, Wireless LAN Research Labs at Worcester Polytechnic Institute has developed a "PlaceTool" application which derives the location of access points from a CAD drawing (as an AutoCAD .dxf file), a technical description of the radio, and radio propagation simulations. The simulations include both statistical analysis and two-dimensional ray-tracing. Barring detailed pre-installation analysis, a good rule of thumb is to assume the need for one access point for each 50,000 square feet of coverage, but this can vary widely based on building construction (particularly with respect to interior walls), prevailing RF conditions, user traffic loads, and the nature and performance of the specific product being used.

2. WIRELESS LAN TECHNOLOGIES

Bandwidth and Regulatory Issues

Most wireless LAN systems use radio frequency (RF) technology, typically in the microwave bands below 6 GHz. The primary reason for this is the availability of spectrum in most parts of the world, most often in the form of unlicensed bands which have no requirement for the licensing of specific users or applications. In the United States, these are known as the Industrial/Scientific/Medical (ISM) bands, with the most popular ISM bands for wireless LANs being at 902-928 MHz., 2.4-2.4835 GHz., and 5.725-5.850 GHz. Similar regulations are in place in much of the world, including Europe (under ETSI 300.328 and national rules) and Japan, although in Japan bandwidth availability is significantly limited.

Parts 15.247 (for spread spectrum implementations) and 15.249 (for narrowband) of the FCC rules (47 CFR) allow use of these bands for wireless LANs, provided certain rules with respect to power output (and antenna gain) are met, and these are of course the responsibility of the radio designer. Moreover, wireless LANs operating in the unlicensed bands must accept any interference present in these bands, and cannot cause interference to those above them in the hierarchy of use for the particular band in question. In practice, however, both of these concerns are minor. While significant noise and potential interference can exist in the ISM bands, the duty cycle of interfering devices tends to be limited, and technology for dealing with the nature of the ISM bands has been realized in the form of spread-spectrum radio and higher-level wireless networking protocols.

Spread-Spectrum Communications

The most common technique used in wireless LANs for coping with potential interference is one of two forms of spread-spectrum communications, a wideband transmission technique. Originally developed for use in military environments, spread-spectrum systems are sometimes referred to a low probability of intercept (LPI), low probability of detection (LPD), anti-jam (AJ), or pseudo-noise (PN) systems. Indeed, the primary benefit of spread-spectrum communications when operating outside of the inherently-noisy ISM bands is enhanced signal security and integrity. Within the ISM bands, however, the use of spread-spectrum is primarily mandated to assure that the power from any given transmitter is evenly spread across the band in use, so as to avoid creating point interference sources and thus polluting the spectrum to the point where communications for all users is impaired. In other words, power per Hertz in a spread-spectrum system is lower, often by two orders or magnitude, than a corresponding narrowband signal. While security and integrity are also important aspects of the unlicensed use of spread-spectrum transmission, it is the power-spreading aspects that are overriding.

The two primary forms of spread-spectrum used in wireless LANs are known as *frequency-hopping spread-spectrum* (FHSS, or simply FH) and *direct-sequence-spread-spectrum* (DSSS, or DS). There is often a temptation on the part of vendors to claim that one technique is universally better than the other, but no such claim can really be made. In general, FH systems offer better battery life, because they consume less power, and are more forgiving of interference from point sources. DS usually has the requirement of a class-A power amplifier in order to maintain linearity across a relatively wide frequency band. On the other hand, DS usually offers better throughput, except in the case of multiple high-amplitude point interference sources simultaneously existing within the band (or portion of the band) being used by a DS system.

Frequency Hopping

An FH system involves a closely-synchronized transmitter and receiver pair. The FH system is, in effect, a narrowband transceiver where the center frequency of transmission varies in a pattern known to both transmitter and receiver. Even though the carrier frequency changes, the effect is one of a continuous narrowband channel, with small temporal gaps as the actual hopping takes place. FCC rules for the 902-928 band require a minimum of 50 hopping points, with a typical bandwidth of 500 KHz. per hop. Average channel occupancy must be no more than .4 seconds in any 20-second period. For 2.4 GHz., the specifications are a minimum of 75 hopping points (1 MHz. each), and .4 seconds average channel occupancy every 30 seconds. Note also that an FH system must use the entire ISM band, even hopping points which may be known at runtime to have point interference sources, but a limited form of adaptive or "look-before-leap" hopping is allowed. Transmission errors which occur due to interference, fading, or other RF issues must be dealt with by higher levels of the protocol stack in use, which can involve both error-correction techniques and packet retransmission. In practice, a single packet will fit within the duration of a single hop point.

The FCC has issued a notice of proposed rulemaking (NPRM) to widen the bandwidth of individual hopping points (with a corresponding reduction in amplitude) to up to 5 MHz., thus allowing FH systems to provide performance of up to 10 Mbps, from today's common 1 –2 Mbps.

Direct Sequence

DS modulation involves converting bits in the data stream to wider, redundant bit strings called *chips*. The typical chipping code used in essentially all 1 – 2 Mbps wireless LAN systems in the 11-bit *Barker Sequence*, a good choice because it has excellent autocorrelation characteristics and meets regulatory requirements for minimum processing gain. With approval of standards for 5.5- and 11-Mbps wireless LANs, a more elaborate form of DS spread spectrum known as *Complementary Code Keying (CCK)* is applied. CCK allows the use of a relatively rich set of codes, and also has excellent multipath resistance. The latter can be very important in indoor radio environments, which is the typical venue for wireless LANs.

Narrowband Systems

Some products use narrowband radio and consequently do not use spread-spectrum techniques. This is allowed in the unlicensed bands at power levels below about 1 mW. Since bandwidth is not devoted to spectrum spreading, more throughput can be obtained. However, interference can be an issue given the relatively high power levels utilized by other occupants of the unlicensed bands.

Infrared (IR) Systems

While most wireless LAN systems use RF of one form or another, IR continues to play an important role in the evolution of the genre. IR has one key advantage in that it is unlicensed on a global basis. This means that an IR-based system can be used anywhere in the world without the approval of a regulatory body. On the other hand, IR is blocked by solid objects, such as walls, which makes the implementation of an IR-based microcellular wireless LAN somewhat challenging. Nonetheless, such systems are available; the general requirement is two or more access points per typical room in order to provide good coverage. The need for access points is somewhat mitigated by the use of diffuse (or reflective) IR technology, which does not require a clear line of sight from transmitter to receiver. While this "bouncing" of IR signals can be quite effective, wall color and texture can affect propagation. Directed infrared systems, which do require a clear line of sight between transmitter and receiver, have also not proven popular, largely due to the necessary restrictions in office configuration required to minimize the chances that an IR beam will be blocked by moving people and other objects.

IR has two other key advantages. The first is the low cost of IR components, which, in volume production, could result in very inexpensive products. This has already been seen in the popular *IrDA* standard, which specifies a point-to-point link, normally between two notebook computers or similar devices. The IrDA 1.1 specification allows for transmissions at up to 4 Mbps over a distance of one meter, making this implementation more of a wireless personal-area network (WPAN) than a true multi-user LAN.

Issues

Most of the potential problems with wireless LANs in operation show themselves as degraded throughput, occasionally degrading to zero throughput when signals cannot propagate at all. The indoor radio environment can occasionally be quite challenging. Problems include various forms of fading (often managed via antenna diversity techniques), multipath (often dealt with via RAKE receiver implementations), interference from both intentional and unintentional radiators, and issues related to overall capacity. Assuming a reasonable and effective layout of the access-point microcells, most problems encountered by users relate to RF propagation issues, particularly deep or shadow fades. These can be diagnosed via the signal-strength applets included with most products, and usually corrected simply by moving the mobile computer or antenna. If this does not work, interference may be the cause, and this case can be evaluated via investigation (for example, microwave ovens are frequently a source of interference in the 2.4 GHz. band).

3. WIRELESS LAN STANDARDS

Most of the installed base of current wireless LAN products are based on vendor-proprietary implementations. This state of affairs has occurred largely out of necessity, as changes in regulatory policy which allow for advances such as wireless LANs usually predate technical standards activities by a many years. Whereas the subject of standards with respect to wired network implementations (whether LAN or not) are critical, due to considerations surrounding such issues as physical connector layout, configuration, voltage and current levels, number of conductors, etc.), wireless has no such physical constraints as roadblocks to productization. This is because interoperability between wireless systems in any configuration other than peer-to-peer can be implemented across a standards-based backbone cable (typically Ethernet) interconnecting hubs or access points. Many vendors also view standards, justifiably, as a "leveling of the playing field", making it more difficult to obtain proprietary advantage in the marketplace. Interoperability as the result of a common standard is nonetheless a requirement for many customers, for reasons primarily related to the protection of investment. Indeed, the lack of industry standards has been quoted by analysts as a reason for the slower-than-expected growth of both vertical and (particularly) horizontal markets for wireless LANs. Standards also provide a degree of comfort to less-technically-sophisticated customers, who cannot be expected to understand wireless technology in order to apply it, and allow for lower component costs as economies of scale made possible by the standard come into play.

The IEEE 802.11 Standard

After nearly seven years of work, the IEEE 802.11 standard was approved in June of 1997. Part of the same committee (802) that is responsible for many other major networking standards, 802.11 was an ambitious project begun in 1990, during the early days of the wireless LAN industry. 802.11 is an unusual standard in a very important dimension. While most standards specify either a single or a small number of interface characteristics, 802.11 is designed to be *extensible*. This means that new features can be added to the standard without changing the functionality already defined by it. Both peer-to-peer and microcellular configurations are specified in 802.11.

802.11 has a single medium access control (MAC) layer [see Figure 2]. This

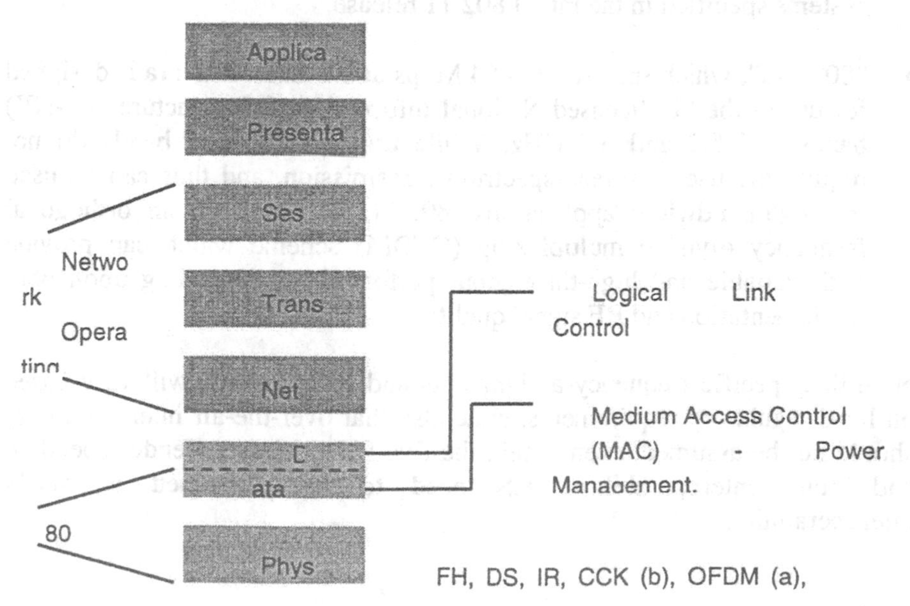

Figure 2 - Wireless LANs and the OSI Model

is the "bottom half" of the *data link* layer, which in turn interfaces with "higher-level" networking functionality, used to share files, printers, and other resources within a computer network. These functions are typically provided by the "network operating system", like Novell's Netware, and Microsoft's Windows 98 and Windows NT. The single MAC definition makes it easier for vendors of this software to adapt their products to work with wireless LANs - only one "driver" is required, and compliance with 802.1D and 802.2 logical link control (LLC) also simplifies the work required to interface an 802.11-compliant wireless LAN. The MAC also provides a common discipline for access the airwaves, based on CSMA/CA, an established multiple-access technique. What is somewhat unusual in 802.11 standard is that the MAC is designed to interface with multiple *physical layers,* or *PHYs*. The PHYS currently defined are as follows:

- FH at 2.4 GHz., with 1 and/or 2 Mbps throughput.
- DS at 2.4 GHz., with 1 and/or 2 Mbps throughput.
- Diffuse IR (not yet implemented commercially).
- "802.11b", sometime called 802.11 HR, which specifies 11, 5.5, 2, and 1 Mbps at 2.4 GHz. The 2 and 1 Mbps rates are compatible with the DS systems specified in the initial 802.11 release.
- "802.11a", which specifies 6 – 54 Mbps at 5.2 GHz. 802.11a is designed for use in the "Unlicensed National Information Infrastructure" (U-NII) bands and 5.2 and 5.7 GHz. While unlicensed, these bands do not require the use of spread-spectrum transmission, and thus can be used for high-bandwidth applications. 802.11a is based on an orthogonal frequency division multiplexing (OFDM) scheme which can provide both variable and high-throughput performance, depending upon both implementation and RF signal quality.

Note that specific frequency assignments and requirements will vary based on local regulatory requirements. Note also that over-the-air interoperability should not be assumed, even within the five PHYs above. Vendor-specified and other interoperability tests need to be performed to verify interoperability.

MAC-Layer Features of 802.11

802.11 is intended to be a robust standard, covering many of the features desirable in wireless-LAN systems. Among these are:

- *Mission Suitability* – Both peer-to-peer and access-point-based configurations are supported.

- *Air Interface* – A variant of CSMA/CA (collision avoidance) known as the *Distributed Coordination Function (DCF)* provides basic access to the airwaves. As collisions could still occur (to say nothing of interference), each packet or fragment of a packet sent is individually ACKed at the MAC layer.

- *Hidden-Node Management* - 802.11 normally uses DCF to avoid collisions. Because a given node might not be able to hear all of the traffic in a given area, it would be logically "hidden" from at least one other node. As a consequence, an optional *RTS/CTS* technique is used as a second layer of protocol to assure orderly access to the airwaves. Since the AP issues the CTS, presumably all stations will hear this and respond accordingly.

- *Asynchronous and Time-Bounded Services* - 802.11 allows for higher-priority messages, via the *Point Coordination Function (PCF)* to receive access to bandwidth ahead of others. This is an informal mechanism, and therefore not rigidly enforceable, but provides a basis for real-time data (for example, multimedia) to be handled efficiently.

- *Power Management* - 802.11 provides a mechanism to allow nodes to "doze" when traffic intended for them is infrequent; nodes "wake up" at predetermined intervals and check to see if any traffic is being held for them by an access point (access points can discard traffic for nodes which don't wake up often enough, for whatever reason). Most mobile wireless systems use some form of "sleep mode" in order to maximize battery life. The length of the period of inactivity required before entering doze mode can be tuned to the requirements of a given application.

- *Multicast Support* - A relatively late addition to the standard is the ability of an implementation to handle both properly-sequenced and out-of-order packets. Multicast packets are not ACKed. Additional work on the standard in this area is underway.

- *Integrity* - 802.11 provides for CRC checking and retransmission when necessary. Acknowledgment (or not) of packet transmissions is provided at the MAC layer, rather than in higher protocol layers, for reasons of timing and efficiency. Messages can also be *fragmented* (broken into smaller units) in noisy situations, allowing a greater likelihood that a given message will get through.

- *Registration, Authentication, and Security* - 802.11 provides for a number of security and privacy mechanisms. Link-level authentication, which assures that a given node is authorized to be on the wireless network, is provided. Specific mechanisms may be defined by individual vendors, but a shared-key encryption technique is specified. A fundamental objective is to make wireless connections as secure as their wired counterparts. This is called *wired equivalent privacy (WEP)* in the standard, and is based on the RSA *RC4* 40-bit algorithm. Many vendors offer more aggressive encryption where locally permitted by law.

WLAN Features Not in 802.11

Even a standard as broad and robust as 802.11 cannot accommodate all functionality needed to implement a complete wireless LAN solution. Some functions were left out because they are part of current or proposed standards (above the MAC layer), and others because agreement could not be reached on a single specification. In addition, some functions that users might assume belong in a wireless-LAN standard were left out because they rightfully belong elsewhere.

Some of the notable missing functionality includes:

- *Access Point Interoperability* - There is no guarantee that access points of different manufacture will be able to handle handoffs. This is because the *wire protocol* used to communicate between access points has not been standardized, and each vendor is free to implement this functionality however they see fit, or to add proprietary extensions such as automatic load balancing or signal-strength optimization. Proposals have been made to correct this (intentional) omission, including the Inter-Access Point Protocol (IAPP) and vendor-specific collaborations.

- *"Layer 3" Functional Mobility* - 802.11 does not address roaming across network boundaries. Such functionality can be provided, for example by "Mobile IP" and "Mobile IPX" solutions provided by others. Note that network-layer functions are not addressed in any way by 802.11; this is in fact a significant feature and minimizes the impact of "going wireless".

- *Network Management* – The role of such management schemes as the *Simple Network Management Protocol (SNMP)* is of great importance in WLANs as, without sophisticated spectrum or network analyzers, RF-related problems can be diagnosed only indirectly. Most products support SNMP capability and this remains a core tool in both overall network management and troubleshooting.

Other WLAN Standards

While 802.11 has captured the greatest degree of attention, the European Telecommunications Standards Institute (ETSI) has been working on high-throughput WLAN technologies under its Broadband Radio Access Networks (BRAN) project. The most notable developments in the area to day include the High-Performance Radio LAN (HIPERLAN) 1 and 2 standards, which have nominal throughput in the 24 Mbps range. HIPERLAN/1 is based on a distributed-control architecture that extends the peer-to-peer model to include forwarding on the order of a distributed router. While the possibility that network backhaul may interfere with user traffic is introduced in this case, the flexibility allowed by the resulting arbitrary mesh structures can be advantageous in many deployment scenarios. Products are expected in 2001.

HIPERLAN/2 has many characteristics in common with 802.11a; both are based on OFDM. It is expected that some harmonization of the two specifications may take place, perhaps resulting in a single unified specification over time. The original mission of HIPERLAN/2 was to be a form of wireless ATM, but the final standard is likely to interoperate with a broad range of wired technologies. HIPERLAN/2 products will probably not appear in the market before 2002.

Bluetooth, a joint venture of the Bluetooth Special Interest Group, is often mentioned as having some characteristics in common with wireless LANs, but it is in fact designed for a different purpose. Bluetooth is a 1 Mbps "wireless personal-area network" (WPAN), designed to allow an individual to wirelessly connect devices over a very short range, nominal a few meters. Examples of the application of Bluetooth include the synchronization of data on two or more personal devices, such as a cell phone or PDA, and using a Bluetooth-equipped cell phone as a relay point between a notebook computer and a wide-area wireless network used to connect to the Internet. Bluetooth does have some LAN-like "piconet" features, but it is not expected to compete with true wireless LANs due to its limited range and throughput.

Finally, standards are being developed for WLANs designed to operate just within the residence. Among the most notable here is HomeRF, which includes wireless IP functionality for data (essentially, a scaled-down version of 802.11), and a variant of the digital enhanced cordless telecommunications (DECT) standard for voice. Thus a single system could handle all residential wireless communications. The goal of systems of this type is, of course, a careful balancing of functionality and cost.

4. DEPLOYING WIRELESS LANS

In most cases, the deployment of wireless LANs is straightforward. Ad-hoc (peer-to-peer) wireless LANs, when operated in a limited physical space with a small (no more than 4-8) number of nodes can be established at will. Infrastructure wireless LANs, based on access points, are more complex, but deployment generally consists of the following a general process:

- An estimate of access point placement is made from a review of layout and construction of the space to be populated. This is only an initial guess, but may be augmented with personal experience with similar wireless LAN products and building structures on the part of the installer.
- Using tools provided by the wireless LAN vendor, a *site survey* is performed. This gives an estimate of coverage and likely performance. Basically, this involves the temporary installation of an access point and

then running a software application on a mobile computer which records throughput characteristics at various points in the area to be covered.

- An estimate of user density is created. This is the number of users per area; many access points have hard limits as to the maximum number of concurrent users, and multiple access points may be required. Multiple access points operating on different radio channels (defined either as physical subchannels in the case of DS, or different hopping sequences in the case of FH) can be used to provide more raw capacity in a given area, and many products support roaming across channel boundaries (i.e., the radio channel of the roaming node is changed by the AP automatically).

- Finally, an examination of likely usage patterns in terms of transmit duty cycles and data rates, and any requirement for isochronous traffic (especially involving large data objects, typical of multimedia applications) needs to be performed.

From this data, a reasonable first pass access-point layout can be created and deployed. After verifying coverage with the site-survey tool, a synthetic benchmark should be run over a reasonably long period (perhaps 24 hours or even more) so as to determine likely real-world performance. A benchmark as simple a file copying can be used provided a good record of transfer volumes and resulting responsiveness is maintained. It is often the case that "fill-in" access points may be required to make up for either gaps in coverage or heavier-than-expected traffic volumes in certain areas of a given installation. Unfortunately, the deployment of most wireless LAN systems involves a degree of uncertainty, but this is easily managed via additional access points in most cases.

Users need to be educated to pay attention to signal-strength tools provided by the wireless LAN vendor [figure 3], which are similar to those on cellular phones. Most throughput-related problems can be solved by simply re-orienting the antenna on the mobile computer, often involving little more than moving the mobile computer a few centimeters one way or the other. If this does not correct the problem, access-point coverage needs to be examined. Problems beyond this are most often the result of interference; microwave ovens and other unlicensed consumer devices (such as cordless phones) are likely suspects. It is very unusual to find pathological problems in wireless LAN throughput – the products are in general reliable, and vendor tools are almost always robust enough to diagnose problems as they occur.

Looking ahead, the future of wireless LANs appears quite bright. Major objections with respect to cost and performance have been addressed, as have issues with size, weight, power consumption, and the potential complications of access-point installation. It appears likely that, with the gradual replacement of desktop computers with mobile systems, wireless LANs will become the connectivity of choice for many users.

Figure 3 - Signal Strength Applet (courtesy 3Com; used with permission)

REFERENCES

Peter T. Davis and Craig R. McGuffin, Wireless Local Area Networks.McGraw-Hill, New York, 1995.

Craig J. Mathias, "Wireless LANs: The Next Wave", Data Communications, March 21, 1992.

Craig J. Mathias, "New LAN Gear Snaps Unseen Desktop Chains". Data Communications, March 21, 1994.

Craig J. Mathias, "Wireless LANs: The Top Ten Challenges". Business Communications Review, August 1994.

Craig J. Mathias, "Wireless: Coming to a LAN Near You". Mobile Computing and Communications, October 1996.

Craig J. Mathias, "Wireless LANs: Getting to Interoperability". Wireless LAN Interoperability Forum; published at www.wlif.com.

Craig J. Mathias, "A Guide to Wireless LAN Standards". Wireless LAN Interoperability Forum; published at www.wlif.com.

Kaveh Pahlavan and Allen Levesque, Wireless Information Networks. John Wiley and Sons, New York, 1995.

Placetool: The Placement Tool Software. Wireless LAN Research Labs, Center for Wireless Information Network Studies (CWINS), Worcester Polytechnic Institute (WPI), Worcester, MA

A. Santamaría and F. J. Lópcx-Hernández, editors, Wireless LAN Systems. Artech House, Boston,1994

Jochen Schiller, Mobile Communications. Addison-Wesley, New York, 2000.

Chapter 12

WIRELESS LANs NETWORK DEPLOYMENT IN PRACTICE

ANAND R. PRASAD, ALBERT EIKELENBOOM, HENRI MOELARD, AD KAMERMAN AND NEELI PRASAD

Wireless Communications and Networking Division, Lucent Technologies, Nieuwegein, The Netherlands

Abstract: Wireless LANs most commonly use the Industrial, Scientific, and Medical (ISM) frequency band, of 2.45 GHz. Although there have been a variety of proprietary solutions, the IEEE approved a standard, 802.11, that organizes this technology. Planning the network, which fulfills the requirements of the user in such systems, is a major issue. In this chapter we will discuss some critical issues faced during wireless LAN deployments from a practical point of view.

1. INTRODUCTION

Proliferation of computers and wireless communication together has brought us to an era of wireless networking. Continual growth of wireless networks is driven by, to name a few, ease to install, flexibility and mobility. These benefits offer gains in efficiency, accuracy and lower business costs. The growth in the market brought forward several proprietary standards for Wireless Local Area Networks (WLANs), this chaos was resolved by harmonizing effort of IEEE with an international standard on WLANs: IEEE 802.11 [1].

Wireless LANs in a Nutshell

Wireless LANs mostly operate using either radio technology or infrared techniques. Each approach has it own attribute, which satisfies different connectivity requirements. Majority of these devices are capable of transmitting information up to several 100 meters in an open environment. In figure 1 a concept of WLAN interfacing with a wired network is given. The components of WLANs consist of a wireless network interface card, often known as station, STA, and a wireless bridge referred to as access point, AP. The AP interface the wireless network with the wired network (e.g. Ethernet LAN) [1], [2], [3].

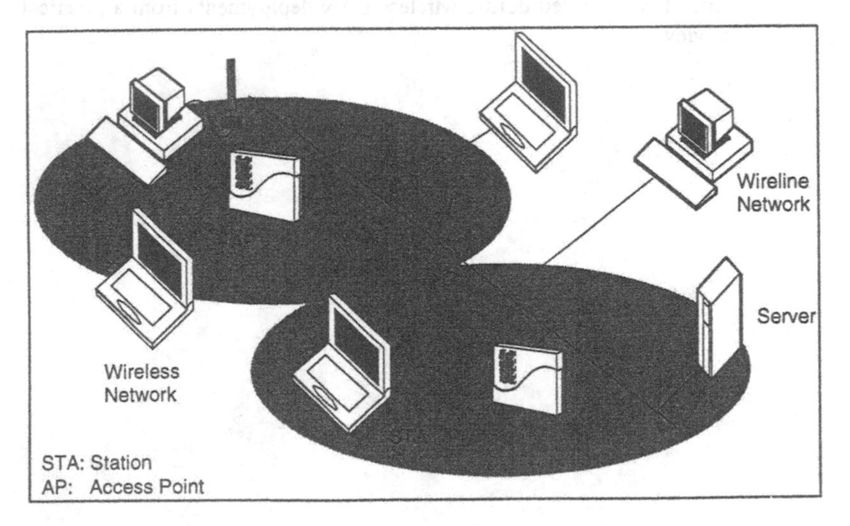

Figure 1. A wireless local area network.

The most widely used WLANs use radio waves at the frequency band of 2.4 GHz known as ISM (industrial, scientific and medical) band. The release of the ISM band meant the availability of unlicensed spectrum and prompted significant interest in the design of WLANs. An advantage of radio waves is that they can provide connectivity for non line of sight situations also. A disadvantage of radio waves is the electromagnetic propagation, which might cause interference with equipment working at the same frequency. Because radio waves propagate through the walls security might also be a problem.

WLANs based on radio waves usually use spread spectrum technology [2], [4], [5]. Spread spectrum *spreads* the signal power over a wide band of frequencies, which makes the data much less susceptible to electrical noise than conventional radio modulation techniques. Spread spectrum modulators use one of the two methods to spread the signal over a wider area: frequency hopping spread spectrum, FHSS, or direct sequence spread spectrum, DSSS.

FHSS works very much as the name implies. It takes the data signal and modulates it with a carrier signal that hops from frequency to frequency as a function of time over a wide band of frequencies. On the other hand direct sequence combines a data signal at a sending STA with a higher data rate bit sequence, thus spreading the signal in the whole frequency band.

Infrared LANs working at 820 nm wavelength provide an alternative to radio wave based WLANs. Although infrared has its benefits it is not suitable for mobile applications due to its line of sight requirement. There are two kinds of infrared LANs, diffused and point to point.

The first WLAN products appeared in the market around 1990, although the concept of WLANs was known for some years. The worldwide release of the ISM band at 2.4 GHz meant the availability of unlicensed spectrum and prompted significant interest in the design of WLANs. The next generation of these WLAN products is implemented on PCMCIA cards (also called PC card) that are used in laptop computers and portable devices[2], [3], [6]. The major technical issues for WLAN systems are size, power consumption, bit rate, aggregate throughput, coverage range and interference robustness.

Considered Wireless LAN

In this chapter we consider WLANs based on DSSS technology as given by IEEE 802.11. The IEEE 802.11 WLAN based on DSSS is initially aimed for the 2.4 GHz band designated for ISM applications as provided by the regulatory bodies world wide [1], [2], [3].

The DSSS system provides a WLAN with 1 Mbit/s, 2 Mbit/s, 5.5 Mbit/s and 11 Mbit/s data payload communication capability. According to the FCC regulations, the DSSS system shall provide a processing gain of at least 10 dB. This shall be accomplished by chipping the baseband signal at 11 MHz with a 11-chip pseudo random, PN, code (Barker sequence).

The DSSS system uses baseband modulations of differential binary phase shift keying (DBPSK) and differential quadrature phase shift keying (DQPSK) to provide the 1 and 2 Mbps data rates, respectively. Complementary code keying (CCK) is used to provide 5.5 and 11 Mbps.

Regulatory Bodies Requirements

The regulatory bodies in each country govern the ISM band. Table 1 lists the available frequency bands and the restrictions to devices which use this band for communications [3], [7]. In the USA, the radiated emissions should also conform to the ANSI uncontrolled radiation emission standards (IEEE Std C95.1-1991).

Table 1. Frequency bands and power levels for wireless LANs.

Location	Regulatory Range	Maximum Output Power
North America	2.400-2.4835 GHz	1000 mW
Europe	2.400-2.4835 GHz	100 mW (EIRP*)
Japan**	2.471-2.497 GHz	10 mW/MHz

* *EIRP= Effective Isotropic Radiated Power.*

** *Japan will adopt the ETS rules as applied in Europe in the year 2000.*

Deployment in General

Scarcity of spectrum is the biggest issue in wireless communication [10]. The challenge is to serve the largest number of users with a specified system quality. For this purpose network deployment and study thereof plays a very important role. In this chapter we will deal with critical issues such as (1) Coverage, (2) Cell planning, (3) Interference, (4) Power management, (5) Data rate and (6) Security especially for IEEE 802.11 WLAN based on DSSS in 2.4 GHz ISM band.

Chapter Organization

We will start this chapter with an explanation on WLAN system design, section 2. In section 3 a study on multiple access scheme (Carrier Sense Multiple Access with Collision Avoidance, CSMA/CA) is given together with results on throughput. A study on RF propagation and coverage is presented in section 4 while interference and coexistence issues are given in section 5. Power management and cell planning are given in section 6 and 7 respectively.

2. SYSTEM DESIGN

In this section we will discuss various aspects of WLAN system design. As systems design can vary, we will concentrate on the system design of Lucent Technologies IEEE 802.11 compliant WLAN system: WaveLAN [2], [3], [6].

Distribution of functions

A WLAN network card and a set of software modules cooperate to offer a 802.11 LAN connection for a PC. At the highest software interface, the equivalent services are offered as for a traditional Ethernet (802.3) LAN. At the air interface, the 802.11 RF/baseband modulation and protocols are used. Figure 2 gives a schematic overview of the major functional elements of the hard- and software and describes how the various functions are distributed over these elements.

A typical WLAN card (in our case Lucent Technologies WaveLAN) is used in laptop computers — the antenna side protrudes from the laptop cabinet. The transceiver front-end is mounted in a plastic cover and this slightly thicker part of the card contains the internal antenna.

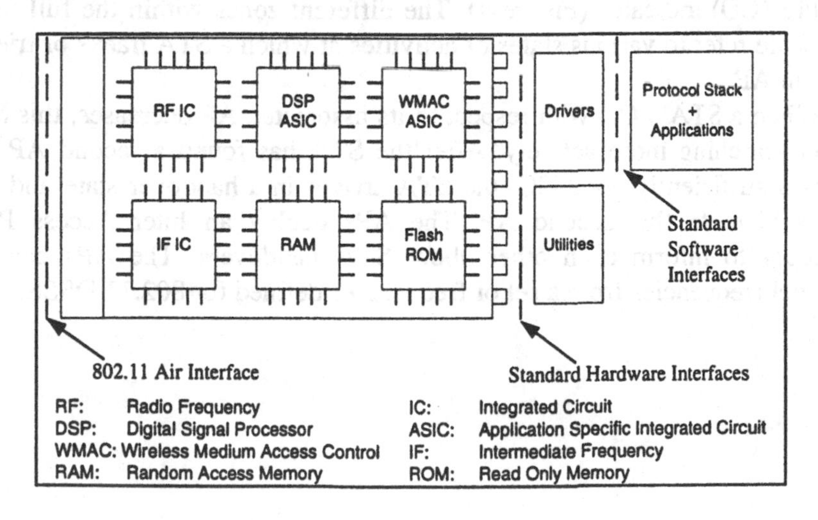

Figure 2. Schematic of a typical WLAN card.

Network cards are available for a number of standard hardware interfaces like PCMCIA, (Figure 3), PCI and ISA.

Figure 3. IEEE 802.11 modem for the 2.4-GHz band (WaveLAN™ from Lucent Technologies).

Roaming

WLANs provide roaming within the coverage boundaries of a set of APs, which are interconnected via a (wired) distribution system. The APs send beacon messages at regular intervals (100 ms). STAs can keep track of the conditions at which the beacons are received per individual AP. The running average of these receive conditions is determined by a communications quality (CQ) indicator (Figure 4). The different zones within the full range CQ scale refer to various states of activities at which a STA tracks or tries to find an AP.

When a STA's CQ with respect to its associated AP decreases, this STA starts searching more actively. After the STA has found a second AP that gives a sufficiently good CQ, the STA arrives in a handover state and will re-associate to this second AP. The APs deploy an Inter Access Point Protocol to inform each other about STA handovers. The APs can use channel frequencies from a set of frequencies defined for 802.11 DSSS.

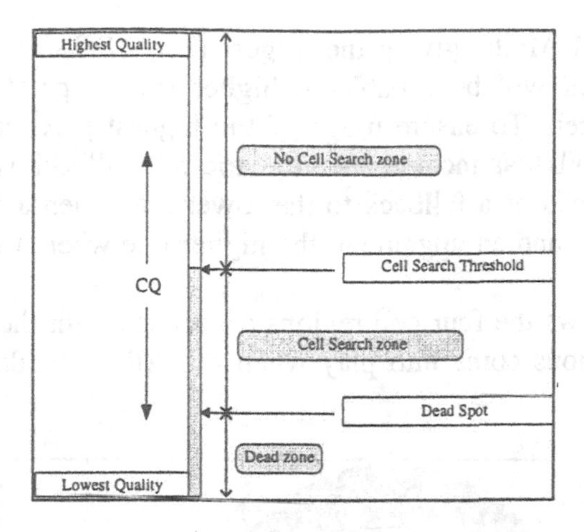

Figure 4. Comms Quality scale and Cell Search zones.

Power management

For battery powered PC devices the power consumption of a LAN card is a critical factor. The 802.11 standard defines power management protocols that can be used by STAs. Power management schemes result in a lower consumption of (battery) power compared to traditional operation where a STA is always monitoring the medium during idle periods. To achieve savings in power consumption, a LAN card in a STA must have a special low power state of operation called DOZE state. In this state the LAN card will not monitor the medium and will be unable to receive a frame. This state differs from the OFF state in the sense that the card must be able to make a transition from DOZE state to fully operational receive (AWAKE) state in a very short time (250 µs). A transition from OFF to AWAKE state will take much more time.

Power Management allows a STA to spend most of its idle time in DOZE state, while still maintaining connection to the rest of the network to receive unsolicited messages. For the latter requirement, the other STAs or the AP must temporarily buffer the messages that are destined to a STA operating in a power management scheme, and such a STA must "wakeup" on regular intervals to check if there are messages buffered for it.

Automatic Rate Fallback

The different modulation techniques used for the different data rates of WLAN can be characterized by more robust communication at the lower rate. This translates into different reliable communication ranges for the different rates, 1 Mbit/s giving the largest range. STAs moving around in such a large cell will be capable of higher speed operation in the inner regions of the cell. To ensure usage of the highest practicable data rate at each moment, WLANs include an automatic rate fallback (ARF) algorithm. This algorithm causes a fallback to the lower rate when a STA wanders to the outer regions and an upgrade to the higher rate when it moves back into the inner region.

Figure 5 shows the four cell regions associated with the four data rates. The ARF functions come into play when the ARF boundary is crossed in either direction.

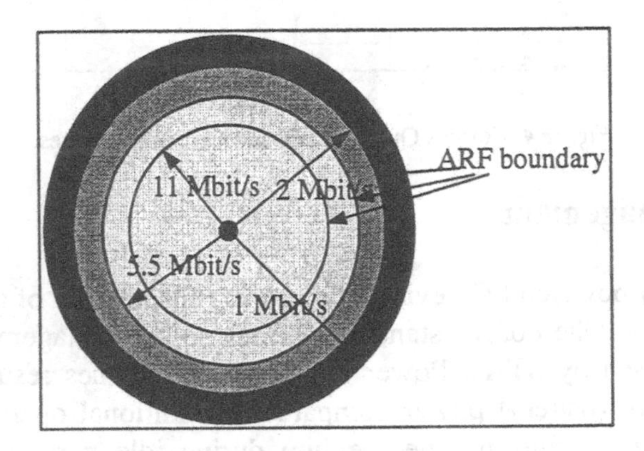

Figure 5. Relation between data rate and cell regions.

Besides resulting in a bigger range, the lower rates will also be more robust against other interfering conditions like high path loss, high background noise, and extreme multipath effects. The ARF scheme will do a (temporary) fallback when such conditions appear and an upgrade when they disappear.

Security

WLANs compliant to IEEE 802.11 combats the security problem with open system and shared key authentication and RC4 based encryption [1], [2], [3], [6], [9]. Open system authentication is essentially a null authentication in which any STA is authenticated by the AP. Shared key authentication supports authentication of STAs as either a member of those who know a shared secret key or a member of those who do not. IEEE 802.11 shared key authentication accomplishes this without the need to transmit the secret key in the clear; requiring the use of the wired equivalent privacy, WEP, mechanism. Closed system authentication, a proprietary scheme, is implemented in WaveLAN, which provides further security. WLAN is envisaged to be used in corporate and public environments and the existing level of security will not be enough for these environments.

In general a corporate environment has an Ethernet based LAN with OS related authentication procedure (Microsoft, Apple, Unix etc.). We will refer to such corporate environment as enterprise environment. Enterprises have closed network environment where reasonable security can be achieved by using network name and shared key authentication. "Reasonable" because shared key and network name based authentication is not a very secure process. Another major concern in Enterprises is the rate of change in personnel, both short and long term. Distributing keys to them and making sure they can not misuse a key once they have left the company is a major managerial problem.

Besides enterprises there are academic and other institutions where either OS based authentication is used or in certain cases Kerberos is used. Kerberos is Unix based; it includes authentication, access control and session encryption. The authentication is decoupled from access control so that resource owners can decide who has access to their resources. In this sense, Kerberos meets the managerial needs given above. For such institutions, the WLAN system must be compatible to Kerberos with the wireless part giving the same level of security as Kerberos.

The public or dial in environment users make use of untrusted communications facilities to access systems of their employer or an internet service provider, ISP. Therefore both authentication and session security are needed. This environment is dominated by Microsoft platforms. Operators and service providers frequently use RADIUS (remote authentication dial in user service). RADIUS services are used especially when people are mobile and require access to their corporate network or when people want to access a ISP from home. WLAN working in such environment will require compatibility to RADIUS and extra security for the wireless part.

3. MEDIUM ACCESS

The 802.11 CSMA/CA protocol is designed to reduce the collision probability between multiple STAs accessing the medium [1], [2], [3], [7], [11]. The highest probability of a collision would occur just after the medium becomes free, following a busy medium. This is because multiple STAs would have been waiting for the medium to become available again. Therefore, a random backoff arrangement is used to resolve medium contention conflicts, Figure 6. A very short duration carrier detect turn-around time is fundamental for this random wait characteristic. The 802.11 standard DSSS uses a slotted random wait behavior based on 20 μs time slots, which cover the carrier detect turn-around time.

In addition, the 802.11 MAC defines an option for medium reservation via RTS/CTS (request-to-send/clear-to-send) polling interaction and point coordination (for time-bounded services).

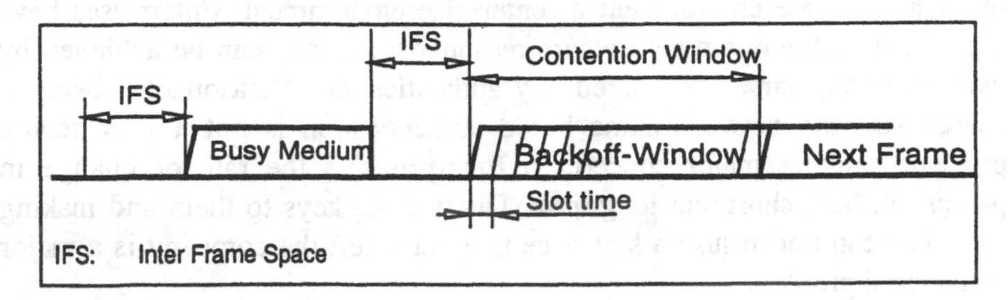

Figure 6. Basic CSMA/CA behavior.

Throughput

Throughput can be measured based on the amount of transferred net data and the required transfer time. A typical method of measuring throughput is by copying a file between a wireless STA and server connected to the wired infrastructure. The effective net throughput depends on the bit rate at which the wireless STA communicates to its AP, but there are a lot of overhead like data frame preamble, MAC (medium access control layer) header, ACK (acknowledgement) frame, transmission protocol overhead (per packet and by request/response packets), processing delay in local/remote computer, forwarding around the AP (Figure 7).

The measurement results are given in Figure 8. To consider throughput measurement with multiple STAs divided over more than one AP we have to look to other aspects like adjacent channel interference (section 5) and medium reuse effects.

4. PROPAGATION AND COVERAGE

The success of any communication system depends on the influence of the propagation medium. Propagation in a medium is affected by atmosphere and terrain [8]. The degree of influence depends primarily on the frequency of the wave. Before proceeding we must understand the propagation characteristics of the frequency assigned for WLANs being studied, 2.4 GHz.

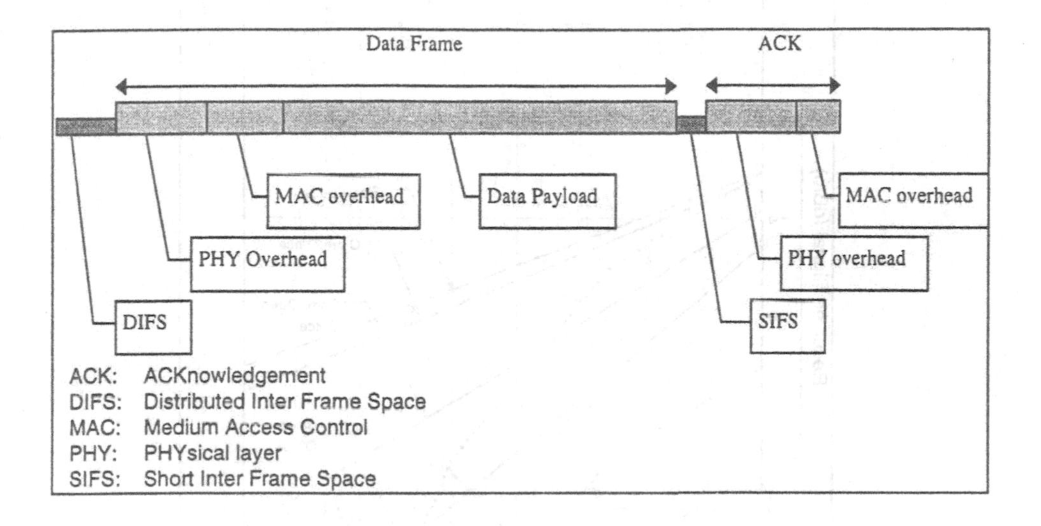

Figure 7. Packet/frame structure of 802.11.

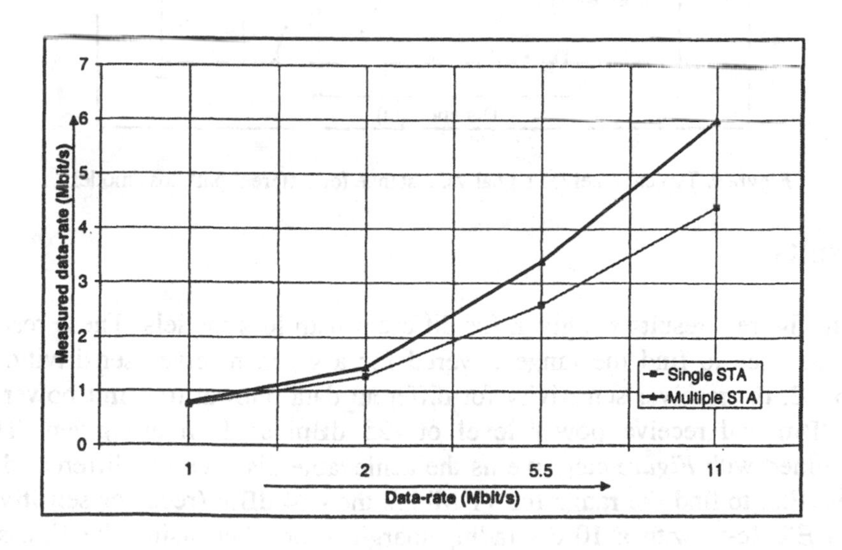

Figure 8. Net throughput (file copy time).

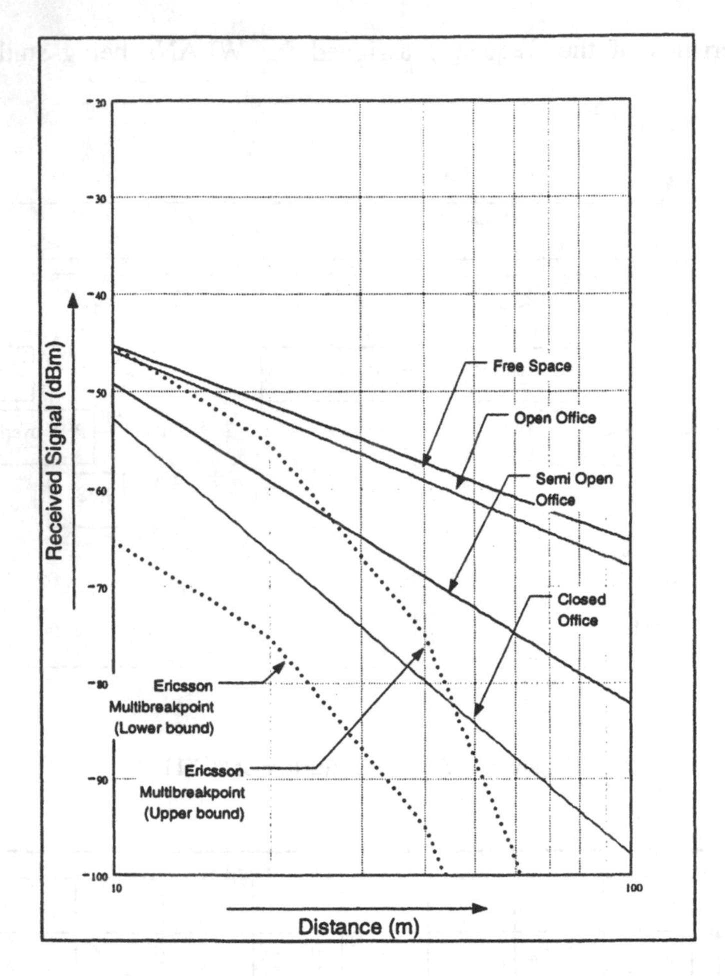

Figure 9. Typical received signal vs. distance for different path loss models.

Results

In figure 9 results are given for different path loss models. These results can be used to find the range covered for a given receiver sensitivity. In Table 2, the receiver sensitivity for different data rates at transmit power of 15 dBm and receive power level of -25 dBm at 1 m are given. This combined with *Figure* can give us the achievable distance for different data rates. E.g. to find the range for 11 Mbit/s the –84 dBm (receiver sensitivity for BER 1e-5) with a 10 dB fading margin (not when using the Ericsson curve) gives –74 dBm, which yields a distance of 29 meters (from the lower solid line (coefficient 4.5).

Experience shows that Ericsson multibreakpoint model gives a very realistic result in environments with obstructions by walls. Realistic situation can vary as the wireless medium is hazardous thus for a given receiver sensitivity the achievable range can vary between the upper and lower bound of Ericsson multibreakpoint model.

Figure 10 gives the receive level, in dBm, around a IEEE 802.11 AP with a 32 mW transmit power installed in semi-open office building. The received signal levels are given for three floors.

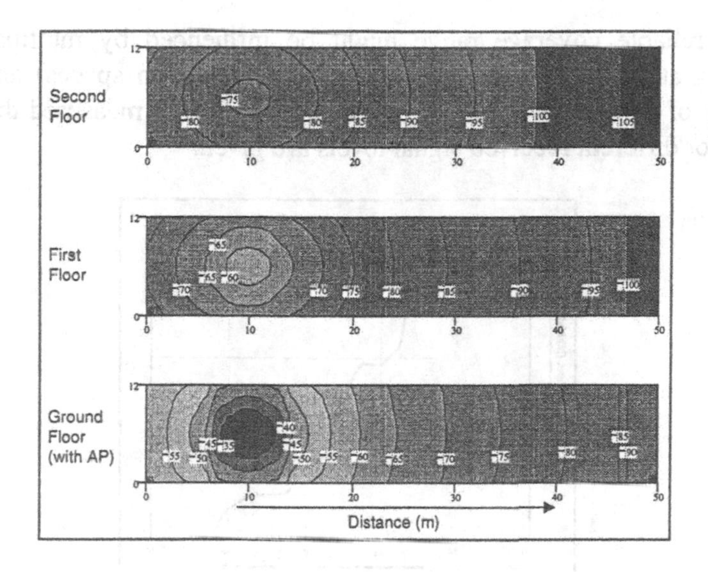

Figure 10. Received signal level (dB) around an IEEE 802.11 AP in an indoor environment.

Coverage

The reliable coverage analysis is based on path loss modeling for environments like Open Plan Building, Semi-Open Office, Closed Office with respective path loss coefficients of 2.2, 3.3 and 4.5 above the 5 meter breakpoint (up to 5 meter free space propagation with path loss coefficient equal to 2). On top of this modeling with path loss dependent on the TX-RX (Transmit-Receive) distance there will be a margin of 10 dB required in relation to variation due to fading. With two antennas and a Rayleigh fading channel the 10 dB margin reflects a reliability of 99%.

Table 2. Reliable range according to path loss models.

Data rate	1 Mbit/s	2 Mbit/s	5.5 Mbit/s	11 Mbit/s
Receiver sensitivity for BER 10^{-5}	-93 dBm	-90 dBm	-87 dBm	-84 dBm
Range covered 99% point TX power 15 dBm				
Open Plan Building (range factor per dB: 1.110)	485 m	354 m	259 m	189 m
Semi Open Office (range factor per dB: 1.072)	105 m	85 m	69 m	56 m
Closed Office (range factor per dB: 1.053)	46 m	40 m	34 m	29 m

The reliable coverage range might be influenced by multipath when operating at 11 Mbit/s and 5.5 Mbit/s (in larger open spaces) and by the presence of concrete walls at all bit rates. In figure 11 measured throughput results for different received signal levels are given.

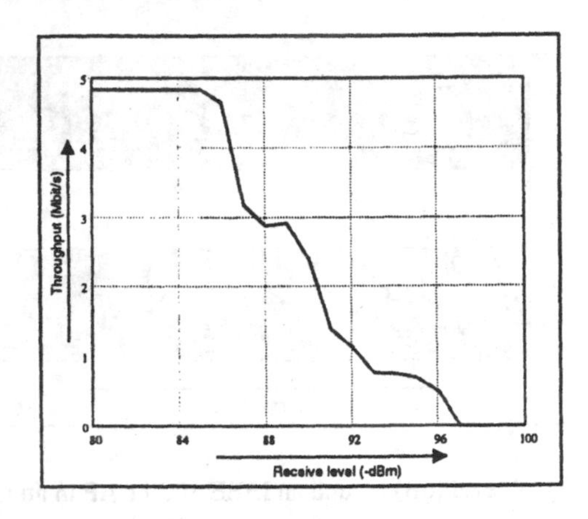

Figure 11. Throughput versus received signal level for 802.11 PC card with ARF.

5. INTERFERENCE AND COEXISTENCE

If a certain frequency band is allocated for a wireless radio system, a fundamental requirement in the efficient use of band is to re-use frequencies at as small a separation as possible [5], [8]. Whenever a band of frequencies is used interference effects have to be taken in account. These can be mainly classified as cochannel and adjacent channel. At the same time in ISM band there is an issue of coexistence with other equipollents working at the same frequency. This section explains and gives results for interference and coexistence for IEEE 802.11 WLAN.

Interference

Radio frequency interference is one of the most important issues to be addressed in the design, operation and maintenance of wireless communications systems. Although both intermodulation and intersymbol interferences also constitute problems to account for in system planning, a wireless radio system designer is mostly concerned with adjacent channel and cochannel interference.

Cochannel interference lies within the bandwidth of the victim receiver and arises principally from the transmitters using the same band. Adjacent channel interference arises from the same sources and causes problems because the receiver filters do not have perfect selectivity.

In the following we talk about adjacent channel interference and microwave interference.

Adjacent Channel Interference

Adjacent cells will not interfere each other when the channel spacing (selected by AP configuration) use channel center frequencies that are 15 MHz separated. With fully overlapping cells the separation has to be 25 MHz to avoid interference and medium sharing. Channel rejection is the combined effect of the transmitter spectrum output shaping, filtering and detection at the receive side. In particular the IF filter (mostly surface acoustic wave, SAW, filter) at the receiver is one of the key components. The required capture ratio (6 dB at 2 Mbit/s, 12 dB at 11 Mbit/s) is fundamental in terms of how robust the scheme is with respect to cochannel interference from neighbor cell that wants to use the same channel, the defer threshold gives the point from where to allow channel reuse.

Defer threshold level and the required capture ratio give the basis of medium reuse planning. The focus could be among others, smaller cells with denser reuse which need more APs, or larger cells to limit the number of APs. At 2 Mbit/s the channel frequency can be reused when there is one other cell in between which is not using that channel frequency.

Figure 12 gives the adjacent channel (and cochannel) signal to interference ratio, SIR, for the considered WLAN system.

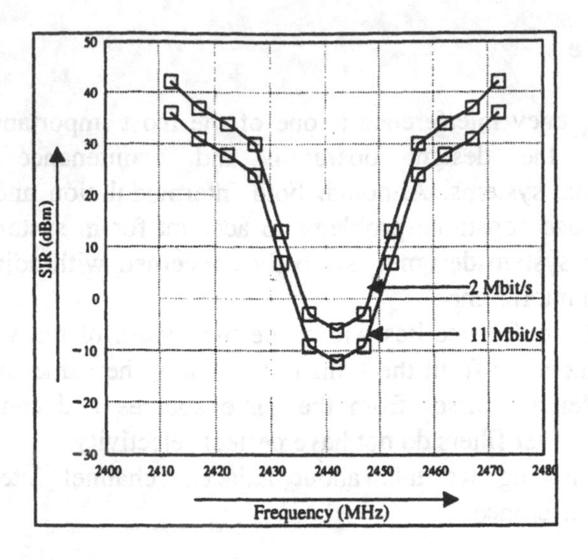

Figure 12. Tolerable adjacent channel interference with center frequency at 2442 MHz.

Microwave Oven Interference

Microwave ovens also work in the ISM band, which creates a lot of noise. Measurement result for WLAN working at 11 Mbps in presence of commercial microwave ovens with AP and STA 3 m apart and microwave oven and STA at 1 m is given as figure 13.

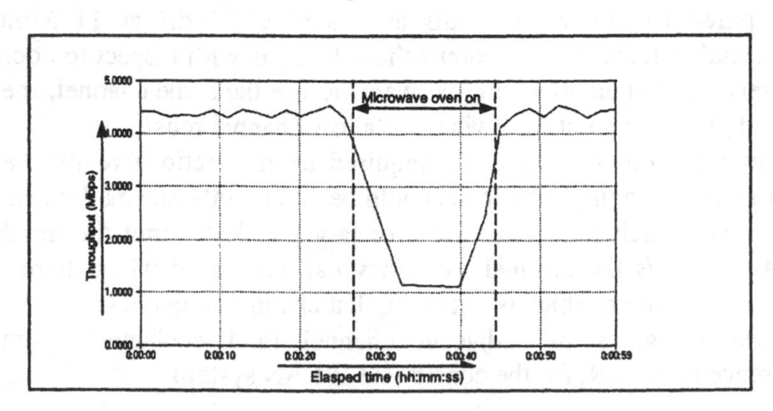

Figure 13. Throughput versus elapsed time for microwave oven interference.

Coexistence

Coexistence is a major issue for wireless communication systems working in ISM band. In this section coexistence study with FHSS IEEE 802.11 and Bluetooth are presented.

FHSS

FHSS 802.11 WLAN stations send one or more data packets at one carrier frequency, hop to another frequency and send one or more packets and continue this hop-transmit sequence (slow frequency hopping). The time these FHSS radios dwell on each frequency is typically fixed at around 20 ms. The FHSS 802.11 uses the modulation technique Gaussian FSK with a low modulation index (Gaussian frequency shaping, BT product = 0.5, modulation index h = 0.34 and 0.15 at 1 and 2 Mbit/s respectively), which gives a relatively narrow spectrum and allows 1 and 2 Mbit/s bit rate in the 1 MHz wide hop bands. However, these FSK conditions result in more sensitivity for noise and other impairments.

Collocated DSSS and FHSS systems interfere with each other in case of channel overlap (11 MHz DSSS channel and the 1 MHz FHSS channel). The tolerable interference level refers to a approximately symmetrical mutual interference situation. DSSS is more robust against in-channel interference because of its despreading (correlation) process. FHSS rejects much of the DSSS signals by its narrower filtering, however, its low-modulation GFSK scheme is much more sensitive to in-channel interference. With single cell DSSS and FHSS systems the channel overlap risk is limited because FHSS hops through the whole 2.45 GHz band. Roughly the tolerable interference for both system in case of channel overlap is 10 dB.

Bluetooth

Bluetooth applies the same modulation scheme as 802.11 FHSS at 1 Mbit/s, however, it hops faster, every 0.625 ms after a period of activity of 0.366 ms and silence of 0.259 ms. The same SIR requirement as in the previous section is applicable, except that the Bluetooth transmit power is 1 mW. Figure 14 shows the SIR with respect to the Bluetooth. System degradation in practice will depend on actual load and traffic process.

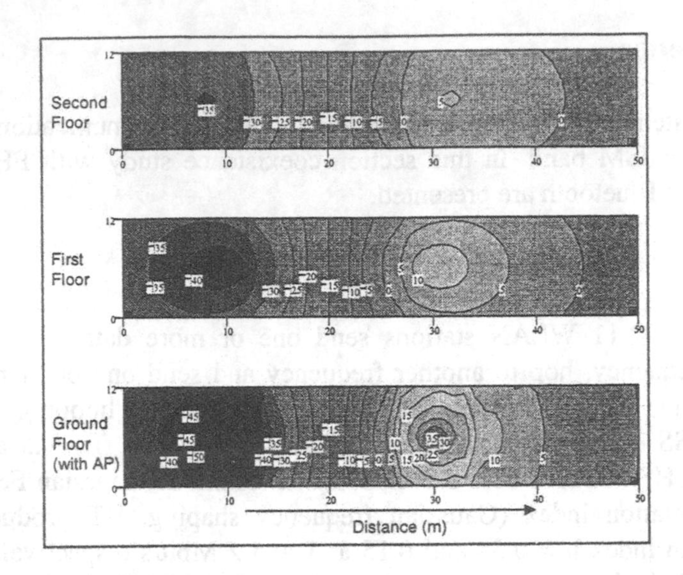

Figure 14. SIR (in dB) for Bluetooth signal with 802.11 interference.

6. IMPACT OF POWER MANAGEMENT

Using Power Management will reduce the amount of current drawn from the battery needed to execute wireless transmissions [3], [6]. The effect of this is an improvement in battery life. Measurement results show that in idle power save mode 15 mA is utilized in contrast to 165 mA in Receive mode and 280 mA in transmit mode.

A system that already consumes a significant amount of battery power for basic components may experience less benefit from power management as opposed to a system where the power needed for the basic platform is low.

On the downside, using Power Management will reduce the overall throughput, as the station has to wake up first to pick up a message that is buffered at the AP.

7. CELL PLANNING AND DEPLOYMENT

There are two basic requirements for a WLAN network deployment: throughput and coverage. On one hand a network can be deployed with the primary requirement being coverage. Such network will have low aggregate throughput and larger cells. Such network deployment will require lesser

APs and thus will be cheaper. On the other hand the primary requirement can be throughput. This means smaller cell size (still giving full coverage), requiring more APs and thus more expensive.

As given in section 0 adjacent cells can be used independently if the center frequencies are spaced 15 MHz apart. With this requirement a frequency planning can be made for each of the different world regions having its own 2.4 – 2.5 GHz ISM band restrictions, Table 3.

Table 3. Number of channels for different world regions

	Channels Available	Number of Channels for Planning
US	2412, 2417, ..., 2462 MHz (11 channels)	4 channels with 15 MHz spacing
Europe	2412, 2417, ..., 2472 MHz (13 channels)	5 channels with 15 MHz spacing
France	2457, 2462, 2467, 2472 MHz (4 channels)	2 channels with 15 MHz spacing
Spain	2457, 2462 MHz (2 channels)	1 channel (with 15 MHz spacing)
Japan*	2484 MHz (1 channel)	1 channel (with 15 MHz spacing)

* *Japan will adopt the ETS rules as applied in Europe in the year 2000.*

The channel reuse distance can be found using the path loss model and the receiver sensitivity as given in section 4. Thus knowing the required data rate and knowing the reusable channel one can deploy the network.

If fully overlapped cells are used center frequencies spaced 25 MHz apart must be used. Of course besides all this the interference sources must be isolated.

REFERENCES

[1] IEEE, "802.11, Wireless LAN Medium Access Control (MAC) and Physical Layer (PHY) specifications," November 1997.

[2] B. Tuch, "Development of WaveLAN, an ISM Wireless LAN", AT&T Technical Journal, vol. 72, no. 4, July/August 1993, pp. 27-37.

[3] A. Kamerman and L. Monteban, "WaveLAN-II: A High-Performance Wireless LAN for the Unlicensed Band", Bell Labs Technical Journal, vol. 2, no. 3, 1997, pp. 118–133.

[4] A. Kamerman, "Spread Spectrum Schemes for Microwave-Frequency WLANs", Microwave Journal, vol. 40, no. 2, February 1997, pp. 80-90.

[5] K. Pahlavan and A.H. Levesque, *Wireless Information Networks*, Wiley, 1995, ISBN 0-471-1067-0

[6] http://www.wavelan.com/

[7] R. van Nee, G. Awater, M. Morikura, H. Takanashi, M. Webster and K. Halford, " New High Rate Wireless LAN Standards", to be published in IEEE Communications Magazine, Dec. 1999.

[8] M.P.M. Hall and L.W. Barclay, *Radiowave Propagation*, IEE Electromagnetic Waves Series 30, London, 1991.

[9] A.R. Prasad, H. Moelard and J. Kruys, "Security Architecture for Wireless LANs: Corporate & Public Environment" under review VTC 2000 Spring, Tokyo, Japan, 15-18 May 2000.

[10] A.R. Prasad, N.R. Prasad, A. Kamerman, H. Moelard and A. Eikelenboom, "Indoor Wireless LANs Deployment", under review VTC 2000 Spring, Tokyo, Japan, 15-18 May 2000.

[11] K.C. Chen, "Medium Access Control of Wireless LANs for Mobile Computing," IEEE Network, September/October 1994, pp. 50-63.

CONTRIBUTORS

PART I OVERVIEW AND ISSUES IN DEPLOYMENTS

Chapter 1 Science, Engineering and Art of Cellular Network Deployment

SALEH FARUQUE received B.Sc. in Physics and MSc. In Applied physics from Dhaka University, Bangladesh in 1969 and 1970 respectively. He received M.A.Sc. and Ph.D degrees in Electrical Engineering from University of Waterloo, Ontario, Canada in 1976 and 1980 respectively. Dr. Faruque had been with Nortel, Networks for the last 18 years and contributed extensively in cellular communications and related areas. At present, Dr. Faruque is a Principal Engineer and Member of the CTO at Metricom Inc., Plano, Texas. In addition to 50+ papers published in various technical journals, he has written a book entitled "Cellular Mobile Systems Engineering", published by Artech House in 1996. He has secured nine patents and has over fifteen patents pending. He is a Senior Member of the IEEE.

Chapter 2 Comparision of Polarization and Space Diversity in Operational Cellular and PCS Systems

*JAY WEITZEN is Professor of Electrical Engineering at University of Massachusetts Lowell. Research interests include wireless communication systems, applications of position location to airborne and terrestrial navigation, and modeling UHF and VHF radio propagation. He is also director of RF Systems Engineering for Nextwave Telecom with responsibility for the RF design of the Nextwave Network and for development of the Tele*Design PCS radio planning tool. Other industrial experience includes GTE Laboratories where he analyzed and characterized cellular system performance, Signatron Technology where he was involved in the design of military VHF systems, and U.S Department of Transportation where he was involved in the analysis of hybrid Loran-C/GPS systems for use in general aviation. Dr. Weitzen received his Ph.D. from the University of Wisconsin, Madison in 1983. He has published over 80 papers in the open literature and is a former editor of the IEEE Transactions on Communication and Senior Member of the IEEE.*

DR. WALLACE is currently with Qualcomm, Inc. and is involved in analysis and standards development for CDMA systems. Previously, at NextWave Telecom, he was involved in various aspects of RF planning and CDMA performance anaysis and prediction for PCS networks. One key activity was the definition and *development of a number of features of NextWave's RF planning tool, TELE*design. He was also involved in RF measurement programs carried out in NextWave's regions, both wideband and narrowband, and development of software for analysis of display of measurement data, and correlation of measurements with the propagation and CDMA performance predictions. Prior to working in cellular communications, Dr. Wallace was Director of Engineering at Signatron, Inc. where he worked on radio systems for communications over fading channels, primarily for troposcatter (beyond line-of-sight microwave) and HF applications. He directed the development of a 10 Mbps digital troposcatter modem, as well as a troposcatter channel simulator. He received his doctorate in electrical engineering from the University of Illinois Urbana-Champaign in 1982.*

Chapter 3 Use of Smart Antennas to Increase Capacity in Cellular and PCS Networks

MICHAEL A. ZHAO has been with Metawave Communications Corporation as a principal engineer since 1997. He has been a technical lead defining and designing the architecture of Metawave SpotLight CDMA smart antenna products. He has also been leading algorithm design, field trial and deployment efforts, and has developed many key techniques and tools for CDMA optimization using smart antennas. From 1994 to 1997, he was with Nortel Technology Ltd., where he worked on CDMA signal processing software and base station transceiver systems, client-server networks, and switching products. Prior to joining Nortel, he had been working on various areas including signal and image processing and remote sensing. He got his BS from the Electrical Engineering Department, University of Science and Technology of China, MS from the Institute of Remote Sensing Applications, Chinese Academy of Sciences, and MAS from the Systems Design Engineering Department, University of Waterloo, in 1982, 1985 and 1991, respectively.

YONGHAI GU, born in Jiangsu Province, China, got his BSc, MSc and Ph.D degrees from the Radio Engineering Department, Southeast University, Nanjing, China, in 1983, 1986 and 1990, respectively. He has more than 19 years of research and product development experience in wireless communications with emphasis on DSP algorithms, modem development and wireless system engineering. He was with Nortel for IS-136 BTS development, Ottawa, Canada, as a DSP engineer, a radio system engineer, and a manager of radio system engineering, in 1993 ~ 1998. Since November of 1998, he has been with Metawave Communications, Redmond, WA, as the GSM smart antenna system architect, now the Sr. manager of the system engineering department. He has published more than 15 papers and holds several patents.

SCOT D. GORDON is a senior network engineer in Metawave Communications Corporation. Gordon plays a key role in the development of smart antenna systems for CDMA, Analog, and GSM cellular networks. His contributions were instrumental to the successful deployment of the first CDMA smart antenna system in a commercial cellular network. Currently, he drives the research and development of the dynamic control functionality for the company's CDMA smart antenna product. Prior to joining Metawave, Gordon was a wireless industry consultant on the design of satellite communications systems. A Ph.D. candidate from the University of Washington, he holds M.S. and B.E. degrees in electrical engineering. His academic research focused on digital communications and statistical signal processing applied to mobile communications, radar and sonar.

MARTIN J. FEUERSTEIN is Vice President of Product Development for Metawave Communications, where he works on the development of smart antenna systems for cellular and PCS networks. Prior to joining Metawave, he was a Technical Manager with Lucent Technologies where he led optimization and applications for cdmaOne systems. Previously he was a Senior Member Technical Staff with U S WEST working on network performance analysis, TIA/IS-95 standards development and experimental field trials of digital wireless systems. Before joining industry, he was a Visiting Assistant Professor with the Mobile and Portable Radio Research Group at Virginia Tech, where his research focused on spread spectrum systems for position location and satellite communications, as well as radio propagation modeling in scattering environments. He has PhD, MS and BE degrees in EE from Virginia Tech, Northwestern and Vanderbilt, respectively.

PART II DEPLOYMENT OF CDMA BASED NETWORKS

Chapter 4 Optimization of Dual Mode CDMA/AMPS Networks

VINCENT O'BYRNE is a Group Manager, Technology Development working on next generation broadband access technologies. He has over 12 years communications experience at GTE working in the wireless, fiber-optic communications, antenna remoting and cable TV areas. For the several years Vincent has worked on GRANET, a tool to optimize the deployment of CDMA over the embedded analog network. As part of this work, Vincent led a team in winning the Warner R&D award in 1997 for "A new set of Tools to Optimize GTE's CDMA Deployment". Prior to that he had undertaken propagation studies in the wireless local loop at PCS frequencies. He has made numerous presentations at conferences and published in refereed professional publications and presently holds six patents. Vincent was intimately involved with GTE's CDMA deployment plans at 900 MHz and helped develop capital planning models (CelCAP) for the PCS effort and worked closely with GTE Wireless on their Strategic planning tool. Vincent has acted as the laboratories wireless technical representative on various international projects in Italy and Germany and worked with GTE International on developing models for the capital cost of deploying various technologies as part of GTE's efforts to seek new opportunities in South America and Asia. He obtained his H.Dip.E.E from Kevin Street College of Technology in 1984, a B.Sc. (Eng.) from Trinity College Dublin (IRL) in 1984, his M.Sc. from Essex University (UK) in 1985 and his Ph.D. from the University of North Wales (UK) in 1988, all in Electrical Engineering, with his doctorate specializing in optical communications. He is presently studying for an MBA at Babson College in Wellesley Mass.

DR. HARIS STELLAKIS is currently a principal member of technical staff at GTE Laboratories Inc., Waltham, MA, specializing on wireless communications. Since October 1997, he has been with the Wireless Access Technologies Department working on technology identification, evaluation and network planning for emerging over-the-air telecommunications applications. From April 1994 to October 1997, he was with the Network Planning and Engineering Department working on the design of CDMA module for the GRANET radio network-planning tool. Before joining GTE, he was affiliated with Motorola Cambridge Research Center, Cambridge, MA, where he conducted R&D on the performance evaluation of digital signal processing applications on Motorola-MIT data-flow high performance computer prototypes. Dr. H. Stellakis received his Ph.D. and Master's degrees from Electrical and Computer Engineering Dept., Northeastern University, Boston, MA, in 1993 and 1989 respectively, and the Diploma in Electrical Engineering from National Technical University of Athens, Greece in 1987. His interests include the technology evaluation and strategy development for the provision of personal communication services and products; analysis, design and implementation of value-added services and products for 2nd and 3rd generation of cellular systems; network planning, deployment and optimization of broadband wireless systems. Dr. H. Stellakis has been a recipient of 1997 Leslie H. Warner award for highest technical achievements in CDMA modeling. He is a member of IEEE Communications, Computer, and Signal Processing Societies; and Technical Chamber of Greece.

RAJAMANI GANESH

His biography is in the section "About the Editors"

Chapter 5 Microcell Engineering in CDMA Networks

JIN YANG received the B.Sc. (Honors) and Ph.D. degrees from Tsinghua University in 1985 and 1989, respectively. Dr. Yang is a Principal Engineer with Vodafone

AirTouch Plc.. She is in charge of AirTouch radio network evolution, design and optimization. She has been invited to speak at major CDMA World Congresses and lecture numerous courses on CDMA network engineering. She has played a key role in the first cdmaOne network commercialization in United States and has led the development of cdmaOne design and planning tool. Before joining AirTouch, Dr. Yang was an adjunct professor at Portland State University teaching wireless communications. She worked on CDMA system development at Sharp Microelectronics Technology Inc. in 1992-1995, TDMA development at NovAtel Communications Ltd. in 1990-1992. She involved in establishing IS-95 CDMA standards in 1992-1993. She has two patents and more than 20 academic papers in wireless communications. Her current interests include CDMA network planning and optimization, CDMA system capacity and performance, wireless communications and telephony.

Chapter 6 Intermodulation Distortion in IS-95 CDMA Handset Transceivers

DR. GRAY earned his B.S. with high honors (1985) and M.S. (1986) in Electrical Engineering from Texas A&M University, and his Ph.D. (1995) in Electrical Engineering from Northeastern University. From 1986 to 1996 Dr. Gray held positions with Sandia National Labs, E-Systems and The MITRE Corporation researching multiuser communications systems, detection for arms control verification and adaptive beam forming. In 1996, he joined Nokia Mobile Phones in San Diego, CA and later became the DSP Algorithms Team Leader for a group responsible for producing the baseband design for Nokia's first IS-95 CDMA handset. Since 1997, Dr. Gray has been a Principal Scientist with Nokia Research Center in Irving, TX where he lead a project responsible for Nokia's system concept for 3rd Generation IS-95 CDMA and recently has been responsible for building a work program in wireless local areas networks. Dr. Gray is a member of Eta Kappa Nu and Tau Beta Pi.

GIRIDHAR D. MANDYAM received the BSEE degree (Magna Cum Laude) from Southern Methodist University in 1989, the MSEE degree from the University of Southern California in 1993, and the Ph.D. degree in electrical engineering from the University of New Mexico in 1996. He was employed with Rockwell International (Dallas, Texas) from 1989 - 1991. He then worked as a teaching assistant at the University of Southern California (Los Angeles, California) from 1991 - 1993. He was employed at Qualcomm International (San Diego, California) from 1993 - 1994, when he took a leave of absence to pursue graduate studies at the University of New Mexico. After completing his Ph.D., he worked at Texas Instruments (Dallas, Texas) from 1996 - 1998. In 1998, he joined Nokia Research Center (Dallas, Texas) where he is currently a Program Manager. He has authored or co-authored over 30 journal and conference publications and three book chapters.

PART III DEPLOYMENT OF TDMA BASED NETWORKS

Chapter 7 Hierarchical TDMA Cellular Network With Distributed Coverage For High Traffic Capacity

JÉRÔME BROUET graduated in 1994 in Telecommunication Engineering from the Ecole Nationale Supérieure des Télécommunications in Paris. He joined advanced studies team in Alcatel Mobile Communication France in November 1995. Since June 1996, he has been working in the Radio Department of the Alcatel Corporate Research Centre in Nanterre on the enhancement of the GSM base sub system (resource allocation for enhanced capacity, GPRS, EDGE) and on wireless data transmission (cellular, WLAN, etc). He is currently team leader and is involved in the design of radio algorithms (base band, MAC and RLC) for EDGE. In addition to his activities in Alcatel, he is a visiting professor in five universities / telecommunication engineering schools in France. He contributed in more than 10 papers in the areas of resource allocation for cellular voice and data services.

DR. ING VINOD KUMAR has a Master and a Ph.D. degree from the Ecole Nationale Sup. de Telecom de Paris. He has more than twenty years of experience in research, development and technical management activities. After working in radio, signal processing algorithm design for Radar and wireline data transmission fields in three different companies, he joined the GSM Cellular Division of Alcatel in France in 1988. He has held positions of Team Leader, Project Manager and is presently Deputy Director of the Radio-communication department in the Alcatel Corporate Research Centre which has teams spread over four locations in Europe. He has been involved in the design of cellular systems of present and future generations (i.e., GSM / DECT / GPRS/ EDGE/ UMTS and HIPERLAN). His main contributions have been the physical layer and MAC layer design and capacity optimization. He has actively contributed in GSM, UMTS and the PCS (in the US) standardization activities. He was on the Evaluation Panel of the RACE-ATDMA Project and has participated in the setting up phase of UMTS Forum. In addition to his activities in Alcatel, he is a visiting professor in six universities / Telecom schools in France. He has contributed to more than twenty journal and conference papers and holds more than ten patents.

ARMELLE WAUTIER received the Dipl. Eng. degree from the Ecole Supérieure d'Electricité, France, in 1986 and Ph. D. degrees from the University of Paris XI,

France, in 1992. Since 1986, she has worked in the Radio department of the Ecole Supérieure d'Electricité, where she is a professor in charge of lectures in the areas of digital communications and mobile radio systems. Her research interests include coding, modulation, multiple access techniques, receiver signal processing, channel modeling, and also radio resource management with applications to mobile radio communications. She participated to European RACE research programs and to studies on GSM, Hiperlan2, and UMTS. She has published journal and conference papers in the areas of channel estimation, equalization and radio access, and she also holds seven patents.

Chapter 8 Traffic Analysis of Partially Overlaid AMPS/ANSI-136 Systems

R. RAMESH is a member of Ericsson Research in Research Triangle Park, NC, where he works on various aspects of cellular and wireless communications. He was previously employed by General Electric Company, where he worked on problems related to Land Mobile Radio. He holds degrees in Electrical Engineering from the Indian Institute of Technology and the California Institute of Technology.

KUMAR BALACHANDRAN completed his baccalaureate with Honours in Electronics and Communications Engineering from the Regional Engineering College, Tiruchi, India in 1986, and his Masters and Doctoral degrees in Computer and Systems Engineering from Rensselaer Polytechnic Institute, Troy, NY, U.S.A, in 1988 and 1992 respectively. His graduate studies focused on convolutional codes and coded modulation with trellis codes. In 1992, he joined Pacific Communication Sciences, Inc., a future subsidiary of Cirrus Logic, where he worked on IS-54 and on the design of the Cellular Digital Packet Data specification. Since 1995, he has been a part of Ericsson's global research organization, is based at Research Triangle Park, North Carolina, and has worked on a range of research problems relating to satellite communications and wireless networks. His current interests lie in Voice over IP for wireless networks. Kumar is the author of a dozen publications, and holds sixteen patents.

Chapter 9 Practical Deployment of Frequency Hopping in GSM Networks for capacity enhancement

DR. ANWAR BAJWA founded Camber Systemics Limited (CSL) in April 1997 and is the managing director of the company. CSL is a wireless technology company that offers strategic consulting and is engaged in developing leading-edge tools for the performance optimisation of GSM, CDMA and 3G UMTS systems. CSL is engaged in the planning of 3G(W-CDMA) in the UK using shared-infrastructure network architecture. Dr. Bajwa was board director responsible for Corporate Technology at MSI from 1991 to March 1997. Prior to that he was Senior Manager at Mercury

One2One and responsible for the planning of the DCS1800 Network in the UK. Between 1986 and 1990 he was Head of Cellular Engineering at Cellnet in the UK. He was also a member of the ETSI GSM2 from 1986 to 1990. Over 25 years and Dr. Bajwa has held various academic posts at British Universities. He is currently a Visiting Industrial Fellow at Bristol University. He has published widely in international technical and research journals.

PART IV DEPLOYMENT OF WIRELESS DATA NETWORKS

Chapter 10 General Packet Radio Service (GPRS)

HAKAN INANOGLU received his Ph.D. in Electrical Engineering from Technical University of Istanbul in Istanbul, Turkey in 1998. He started to work for Military Electronic Company of Turkey (ASELSAN) as a RF Design Engineer from 1987 to 1991. From 1992 to 1996 he acted as a Senior Wireless System Engineer for Nortel in Istanbul, Turkey where he worked in various wireless projects including DECT, GSM as a system engineer. He represented Nortel (Istanbul) in ETSI RES03 meetings for DECT standardization. He joined Omnipoint Technologies (OTI) in 1996. He has worked in radio performance, power control, deployment /capacity and coexistence simulation and analyses of US PCS technology IS-661. After OTI split up he joined to Opuswave Networks as a Staff System Engineer. He is currently working as a wireless IP system designer for a C-GSM product. Dr. Inanoglu has presented several articles in various conferences about propagation modeling, deployment and wireless systems.

JOHN K. REECE received the BSEE degree in Electrical Engineering from Wichita State University, Wichita, KS, in 1980. From 1980 to 1984 he was a member of the defensive avionics group at the Boeing Military Airplane Company in Wichita, KS, where he was involved in the design, installation, and measurement of wideband phased array and fixed beam antennas for the B-52 Electronics Countermeasures Systems. From 1984 to 1992 he worked on advanced spacecraft antenna and system designs with the Martin Marietta Astronautics Group in Denver, CO. In 1992 he formed a company to do research contracts for the U.S. defense department looking at advanced phased array antenna technology for shipboard and air-to-air intercept missile communication and radar tracking systems. In 1994 he joined Omnipoint Corporation where he has been involved in the design of specialized antenna systems for personal communication systems, radio system engineering analysis, and propagation predication, analysis, and measurement. Mr Reece holds two U.S. Patents for personal communications antenna designs.

MURAT BILGIC received his Ph.D. in Electrical Engineering from Concordia University in Montreal, Canada in 1992. He has worked as a Senior System Engineer for Nortel in Istanbul, Turkey until the end of 1994. During his tenure, Dr. Bilgic has participated in the national deployment of SS7 in Turkey. He joined Omnipoint Technologies (OTI) in the beginning of 1995. He has worked extensively in the design and development of US PCS technology IS-661. He represented Omnipoint in various standardization forms, such as T1S1, ETSI GPRS ad-hoc, etc. He is currently working as a Principal Engineer providing consulting services to clients of OTI in the field of GPRS. Dr. Bilgic has published more than ten articles in journals and refereed conferences about protocol engineering, multimedia networks, and GPRS.

Chapter 11 Wireless LAN Deployments: An Overview

CRAIG J. MATHIAS is a Principal with Farpoint Group, an advisory and systems-integration firm in Ashland, MA. Farpoint Group specializes in emerging communications technologies, products, and services, with an emphasis on wireless data. The company works with both manufacturers and end-users in technology assessment, strategy development, product specification and design, product marketing, program management, and business-process re-engineering, across a broad range of markets and applications. Craig holds an Sc.B. degree in Applied Mathematics and Computer Science from Brown University. He has published numerous technical and overview articles on a number of topics, and is a well-known industry analyst and frequent speaker at conferences and trade shows. He is also a member of the Advisory Board for the COMDEX Fall conference, and a member of the IEEE.

Chapter 12 Wireless LANs Network Deployment in Practice

ANAND RAGHAWA PRASAD received M.Sc. degree in Electrical Engineering from Delft University of Technology, The Netherlands, in the field of "Self Similarity of ATM Network Traffic" in 1996. From 1996 to 1998 November he worked as Research Engineer and later Project Leader in Uniden Corporation, Tokyo, Japan. Since 1998 December he has been working as Systems Architect for Wireless LANs in Lucent Technologies, Nieuwegein, The Netherlands. He will complete his Ph.D. by end 2000 from Norwegian University of Science and Technology, Trondheim, Norway. The tentative title of the thesis is Indoor Wireless LANs: Protocols, Security and Deployment. He has published several papers in journals and international conferences. His research interests lie in the fields of security and QoS for WLANs, wireless and mobile internet access and software radio.

J. ALBERT EIKELENBOOM was born in Honselersdijk, The Netherlands, in 1969. He received his M.S. in Electrical Engineering From Delft University of Technology, with as specialisation radio communication. After a short period of working at Technion, the University of Haifa, Israel, he joined AT&T in 1995 to work as RF design engineer in the 2.4 GHz band. Currently he works with Lucent Technologies The Netherlands as a system engineer for the WaveLAN project.

HENRI MOELARD holds B.Sc. in Electrical Engineering, Computer Technology, and has over 20 years experience in data communication software and systems developments with NCR Corporation, AT&T, and Lucent Technologies. Since 1990 he is working as Systems Architect for WaveLAN. He helped develop the IEEE 802.11 MAC protocol and standard, as well as the Inter Access Point Protocol.

AD KAMERMAN (1954) is a member of technical staff assigned to a product development team in the Wireless Communication and Networking Division of Lucent Technologies in The Netherlands. He works on wireless LANs, designing their system performance as it relates to the radio-frequency environment, signal processing, and network throughput. Mr. Kamerman received a B.S. and M.S. in electrical engineering from Twente University of Technology in The Netherlands.

NEELI R. PRASAD received M.Sc. degree in Electrical Engineering from Delft University of Technology, The Netherlands, in the field of "Indoor Wireless Communications using Slotted ISMA Protocols" in 1997. She joined Libertel, Maastricht, The Netherlands as a Radio Engineer in 1997. Since November 1998, she is working as Systems Architect at Lucent Technologies, Nieuwegein, The Netherlands. She has published several papers in international conferences. In December 1997 she won Best Paper award for her work on ISMA Protocol (Inhibit Sense Multiple Access). Her current research interest lies in wireless networks, packet communications, multiple access protocols and multimedia communications.

About the Editors

RAJAMANI GANESH received the B.E. degree from Indian Institute of Science, Bangalore, India and the Ph.D. degree in Electrical Engineering from Worcester Polytechnic Institute, Worcester, Massachusetts. From 1991 to 1995 he was part of the Communications Research Laboratory at Sarnoff Corporation in Princeton, New Jersey. While at Sarnoff, he worked on several communication systems design projects including the HDTV transmission system and digital cellular CDMA networks. Since 1995, he has been a part of the Network Planning and Engineering department at GTE Laboratories near Boston. He is actively involved in wireless system design analysis and research and his work has been incorporated in GRANET - a cellular radio planning tool for network optimization which helps GTE Wireless and other GTE business units optimally plan their analog and digital cellular radio networks. He has published several technical papers in many journals and conferences, holds and has applied for many patents and has received numerous awards including the prestigious WARNER award, GTE's highest award for outstanding technical achievement. Dr. Ganesh has been an organizing and technical program committee member of many international conferences and is a technical program chairman of the International Conference on Personal Wireless Communications, to be held in December 2000 in India. He was also the lead editor of the book "Wireless Multimedia Network Technologies", published by Kluwer Academic Publishers in 1999.

KAVEH PAHLAVAN, is a Professor of ECE, a Professor of CS, and Director of the Center for Wireless Information Network Studies, Worcester Polytechnic Institute, Worcester, MA. His area of research is broadband wireless indoor networks. He has contributed to more than 200 technical papers and presentations in numerous countries. His is the principal author of the Wireless Information Networks, John Wiley and Sons, 1995. He has been a consultant to a number companies including CNR Inc, GTE Laboratories, Steinbrecher Corp., Simplex, Mercury Computers, WINDATA, SieraComm, and Codex/Motorola in Massachusetts; JPL, Savi Technologies, RadioLAN in California, Airnoet in Ohio, United Technology Research Center in Connecticut, Honeywell in Arizona; Nokia, LK-Products, Elektrobit, TEKES, and Finnish Academy in Finland, and NTT in Japan. Before joining WPI, he was the director of advanced development at Infinite Inc., Andover, Mass. working on data communications. He started his career as an assistant Professor at Northeastern University, Boston, MA. He is the Editor-in-Chief of the International Journal on Wireless Information Networks. He was the program chairman and organizer of the IEEE

Wireless LAN Workshop, Worcester, in 1991 and 1996 and the organizer and the technical program chairman of the IEEE International Symposium on Personal, Indoor, and Mobile Radio Communications, Boston, MA, 1992 and 1998. He has also been selected as a member of the Committee on Evolution of Untethered Communication, US National Research Council, 1997 and has lead the US review team for the Finnish R&D Programs in Electronic and Telecommunication in 1999. For his contributions to the wireless networks he was the Westin Hadden Professor of Electrical and Computer Engineering at WPI during 1993-1996, was elected as a fellow of the IEEE in 1996 and become a fellow of Nokia in 1999. From May of 2000 he will be the first Nokia-Fulbright scholar and International Professor of the Telecommunication Laboratory, University of Oulu, Finland.

Index

N

O

P

Q

R

S

T

V

W